ENVIRONMENTAL POLICY ANALYSIS FOR DECISION MAKING

THE ECONOMICS OF NON-MARKET GOODS AND RESOURCES

VOLUME 1

Series Editor: Dr. Ian J. Bateman

Dr. Ian J. Bateman is Professor of Environmental Economics at the School of Environmental Sciences, University of East Anglia (UEA), UK and Senior Fellow at both the Centre for Social and Economic Research on the Global Environment (CSERGE) and the Centre for the Economic and Behavioural Analysis of Risk and Decision (CEBARD), UEA. Professor Bateman is also Executive Editor of *Environmental and Resources Economics*, a journal published in cooperation with the European Association for Environmental and Resource Economists (EAERE).

Aims and Scope

The volumes which comprise *The Economics of Non-Market Goods and Resources* series have been specially commissioned to bring a new perspective to the greatest economic challenge facing society in the 21st Century; the successful incorporation of non-market goods within economic decision making. Only by addressing the complexity of the underlying issues raised by such a task can society hope to redirect global economies onto paths of sustainable development. To this end the series combines and contrasts perspectives from environmental, ecological and resource economics and contains a variety of volumes which will appeal to students, researchers, and decision makers at a range of expertise levels. The series will initially address two themes, the first examining the ways in which economists assess the value of non-market goods, the second looking at approaches to the sustainable use and management of such goods. These will be supplemented with further texts examining the fundamental theoretical and applied problems raised by public good decision making.

For further information about the series and how to order, please visit our Website
http:/www.wkap.nl/series.htm/ENGO

Environmental Policy Analysis for Decision Making

Edited by

John Loomis

Department of Agricultural and Resource Economics,
Colorado State University,
Colorado, U.S.A.

and

Gloria Helfand

School of Natural Resources and the Environment,
University of Michigan,
Michigan, U.S.A.

SPRINGER-SCIENCE+BUSINESS MEDIA, B.V.

A C.I.P. Catalogue record for this book is available from the Library of Congress.

ISBN 978-90-481-5532-3 ISBN 978-0-306-48023-2 (eBook)
DOI 10.1007/978-0-306-48023-2

Printed on acid-free paper

Contents

PREFACE

1. ENVIRONMENTAL POLICY ANALYSIS: WHAT AND WHY?

Why environmental policy analysis? Environmental issues are growing in visibility in local, national, and world arenas, as a myriad of human activities leads to increased impacts on the natural world. Issues such as climate change, endangered species, wilderness protection, and energy use are regularly on the front pages of newspapers. Governments at all levels are struggling with how to address these issues. Environmental policy analysis is intended to present the environmental and social impacts of policies, in the hope that better decisions will result when people have better information on which to base those decisions. Conducting environmental policy analysis requires people who understand what it is and how to do it. Interpreting it also requires those skills. We hope that this book will increase the abilities, both of analysts and of decision-makers, to understand and interpret the impacts of environmental policies.

Policy analysis books almost invariably begin by pointing out that policy analysis can take many forms. This book is no different. As you will see in Chapter 1, we consider policy analysis to be information provided for the policy process. That information can take many forms, from sophisticated empirical analysis to general theoretical results, from summary statistics to game theoretic strategies. The form of policy analysis that we present here draws from the experiences that each of us has had in conducting policy analyses, and those experiences are shaped by the fact that both of us are economists. We cannot cover all the necessary topics, especially for a subject as large and varied as environmental policy. Nor do we claim that every policy analysis must include all the topics that we present. Still, we hope and expect that this book provides a number of useful concepts and tools that will assist many people involved in policy.

We focus on environmental and resource policy analysis in this book. Probably the most important difference, except in specific application, between the approaches we cover here and the approaches used in a general public policy oriented book is our emphasis on the importance of factors beyond those based in economics and politics. Environmental issues are inherently interdisciplinary, with critically important natural and physical science aspects as well as social science aspects (not to mention occasional forays into ethics). The natural and physical science impacts of environmental policy have to be part of the debate for environmental policies.

Because of the interdisciplinary nature of environmental policy, we emphasize (see Chapter 3) that an environmental policy analysis could (and probably should) include a wide range of information, including (but not limited to) scientific, economic, sociological, and feasibility effects of a policy proposal. A policy analyst may be required to synthesize all that information into a form readily understandable to a decision-maker, or the analyst may actually conduct the studies necessary to develop some of that information. We propose that all the relevant information be arrayed in a matrix. One dimension of the matrix will list the criteria that should be considered in evaluating the decision, including the environmental and social effects. The other dimension lists the alternatives to be evaluated. Presenting each alternative and its effects next to other alternatives and their effects permits ready comparison of the alternatives.

We present a range of decision rules that an analyst may use to determine the best alternative. Note that we have said "a range of rules." Legally, politically, and morally, we find it hard, if not impossible, to argue that there is one decision rule that applies to all environmental policy choices. As economists, we can (and will) argue that an efficiency criterion – maximizing net benefits to society – has many advantages. We also know (and present) its limitations as a decision rule and suggest alternative rules (also in Chapter 3). The decision must, by its nature, take a wide range of factors into account, and the decision-maker's job is specifically to make the final call (or she wouldn't be a decision-maker). The analysis does not, on its own, determine the decision. The analysis may determine the most efficient alternative, or the least environmentally damaging alternative, or the alternative with the most desirable distributional impacts, but it is up to the decision-maker which, if any, of these objectives is the one on which she will base her decision.

Both relating to general principles and at the down-and-dirty data analysis level, we highlight the ethical and practical dilemmas that a policy analyst may face. We believe in conducting a neutral, objective analysis of whatever topic an analyst faces. Doing otherwise is professionally unconscionable and can (and perhaps should) end a career. At the same time, it is hard to choose an approach for analysis that is completely independent of the decision criteria. A benefit-cost analysis suggests the importance of economic efficiency; a distributional analysis emphasizes who gains and who loses; an ecological analysis puts primacy on the impacts on the natural world. Additionally, data or time limitations (see Chapter 2) may force a tradeoff between the timeliness of an analysis and its quality, or analysis may be used to delay a decision. As academics ourselves, we prefer analysis that draws on soundly developed data sets that are designed for the work at hand, that uses tools that are theoretically sound and appropriate for the circumstances, and that permits the time and resources to do the analysis right. We have also both conducted studies where these conditions have not been fully met (ssshhh, don't tell our deans). A policy analyst, to be relevant, has to work in a world of limited data, limited time, and a client who wants a particular piece of information (and may even want a particular answer to a question). While we cannot tell every analyst how to deal with these issues, we do believe that a forewarned analyst is better armed to face them.

The specific tools that we cover in this book – modeling, statistical regression, benefit-cost analysis, nonmarket valuation, input-output analysis, optimization and linear programming, and risk analysis – are ones that, in our experience, can be very valuable to an analyst (and a decision-maker). All of them are commonly used in environmental policy analysis, as our examples suggest, and their results often appear in environmental impact statements. Our aim in these chapters is to provide the reasons for using these tools as well as warnings about how they can be (and often are) misused. If reading these chapters is combined with some guidance, support, and hands-on experience, we hope that they can equip an analyst to use these tools. At the very least, we expect that an analyst (or decision-maker) will be better prepared to read and understand someone else's use of these tools.

2. AUDIENCES FOR THIS BOOK

The origins of this book lie in a course, Natural Resources and Environment (NRE) 471, Analytical Tools for Environmental Policy, that Gloria Helfand has been teaching at the University of Michigan's School of Natural Resources and Environment. This course covers most of the topics you see included in this book. John Loomis has previously written about many of the principles and tools that are covered here (indeed, Helfand has used those materials in her course prior to the availability of this book). Both of us have worked on environmental policy issues for clients (including government agencies and environmental organizations), as well as having conducted academic research in university settings. It thus made sense to us to combine forces to prepare this book.

While NRE 471 provided the original intended audience for this book, we like to think that others will find this book useful. Perhaps the most obvious audience is other students and teachers of environmental policy analysis. While other public policy analysis books exist, we are unaware of any that focus on environmental issues, and students may appreciate examples in their areas of interest. Despite that last sentence, it is also possible that public policy programs can find a use for this book, either in specific environmental classes or in general-purpose courses. Even though the examples here are all environmental, many of the principles and tools do not change if the subject becomes health care, tax policy, or space flight. Finally, we hope that practitioners in the field – both policy analysts and decision-makers on environmental topics – will learn something from reading this book. We believe (as we've said before and as we will say again) that better information makes for better decisions, and policy analysts can play a very important role in providing better information.

Those using this book – in particular, the tools chapters – will probably find it easier if they have some previous background in economics and statistics. At the same time, we have sought to provide sufficient explanation that an educated non-specialist with an interest in this material will be able to understand the techniques. Additionally, because we have sought to emphasize general principles, even in the tools chapters, students with advanced skills in some of these topics are likely to gain from the broader context we provide for their specialized methods. Our ultimate dream is that everyone who conducts or uses environmental policy analysis will have a well-thumbed copy of this book at hand at all times. We leave up to you whether that dream will have any basis in reality.

In the meantime, whether or not you keep your well-thumbed copy available at all times, we hope you will provide us with feedback on this book. NRE 471 has improved over the years it has been taught because of feedback from those who took it before. We both have been involved with environmental policy analysis because we wanted to provide useful information for the policy process. We hope that this book is a continuation of those efforts, by helping others provide and use information. Please feel free to send us photocopies of any pages you feel have errors or confusing writing or examples. We also welcome any particularly good (or bad) environmental policy analyses that you think would be useful in any future editions of this book, or suggestions for new topics, or feedback on how we have presented our materials.

3. ACKNOWLEDGEMENTS

The students who have passed through NRE 471 served as Helfand's inspiration for this text. Over the years, they have provided valuable feedback on early versions of some of these chapters and encouragement on this project overall. You know who you are, and I thank you for the fun of participating in the class with you.

Kate Cardamone and Janelle Kinkor, both at the time master's students in both the School of Natural Resources and Environment and the Ford School of Public Policy at the University of Michigan, found and wrote up many of the examples that appear in the text. They also read and made suggestions on early versions of many chapters and provided much good humor. They are absolved of all blame (but deserve positive credit) for what shows up here. Several of the empirical examples in this book draw upon research with my former graduate students Michelle Haefele, Robert Richardson, Sabina Shaikh and Pierre duVair as well as Professor Stephan Weiler and Marvin Feldman of Resource Decisions. We also owe a debt of gratitude to Ben Aldridge at Colorado State University who did a great job proof reading the "camera ready" chapters for consistency between figures, tables, text and references. His keen eyes have saved readers of this book much frustration.

Both of us have had a variety of experiences working on public policy issues, and our thinking on this subject has evolved and expanded from these experiences. We would like to thank the many agencies and organizations that have allowed us these opportunities.

I (Helfand) would like to thank my coauthor (Loomis) for the great fun and education from working on this project with him. I hope writing a book with me has been half as easy as it has been working with you. I (Loomis) wish to thank my coauthor (Helfand) for being delightful to work with. She made a laborious effort less of a task.

Gloria Helfand, School of Natural Resources and the Environment, University of Michigan

John Loomis, Dept. of Agricultural and Resource Economics, Colorado State University.

Chapter 1

What Is Policy Analysis, and What Is Its Role in the Policy Process?

1. WHAT YOU NEED TO KNOW ABOUT POLICY ANALYSIS BEFORE YOU BEGIN TO DO IT

What is policy analysis? One general statement is that policy analysis provides information into a public decision making process, or policy process for short. By policy process we refer to the means by which a governmental body (e.g., an agency or legislature) decides upon a policy. By policy, we mean a course of action of a governmental body, where a course of action could be a law, a regulation, a project, or other public decision. Many people in addition to those who can enact policy are likely to be interested in it, including elected officials, representatives of other government agencies, affected individuals, businesses, and non-governmental organizations. These individuals likely want to understand the impacts of a policy, upon themselves, upon others, and upon the wider world. Knowledge of the impacts of a policy can lead any of these individuals to take or change a position on a policy, or to work to amend the policy. Thus, understanding the effects of a proposed policy can affect what the policy will be, whether the policy will be adopted, and how effective it will be.

This view of the policy process assumes that more information is better than less, and that decision-makers can, will, and should incorporate information into their decisions. A policy analyst therefore has to think about how to make that information useful to a decision-maker.

Why is policy analysis important for environmental issues? Often, "the environment" is considered a topic only for engineers, biologists, chemists, or other physical or natural scientists. For instance, the debate about climate change and the "greenhouse effect" is often framed in terms of the scientific uncertainty surrounding the issue. If scientists cannot agree on how significant a problem climate change may be, then perhaps we should not take action to prevent climate change. Or, if scientists agree that climate change is a problem, then perhaps we should take all actions necessary to prevent it. In contrast, if climate change is viewed as a policy question rather than as a scientific question, the scientific uncertainties and their resolution are only part of the issue. Instead, other issues arise, including: what are the costs of controlling climate change? What are the consequences of not reducing climate change? Who gains from reducing the effects of climate change, and who gains from *not* controlling climate change? Raising these sets of additional questions indicates that the scientific issues underlying climate change may not be the most important ones for development of policy; instead, the consequences of a policy (or of not having a policy) for various interests can have a significant effect on whether legislation is passed or treaties signed. Additionally, it becomes clear that environmental issues are inherently multidisciplinary, involving the expertise not only of physical and natural scientists but also of social scientists and people from the humanities. Finally, it becomes clear that many different issues are involved in the determination of what, if anything, should be done about climate change. Indeed, different people involved in these issues are likely to have interests in different parts of this topic and will focus on different questions.

This chapter will put policy analysis into a broader philosophical context. What kinds of information should a policy analyst provide? What is the role of the client in conducting policy analysis? Should a policy analysis provide a specific recommendation or just present an analysis of the options? If so, on what objective should the recommendation be based? Finally, this chapter will discuss some current federal requirements for policy analyses for regulatory actions. If government agencies now face mandates to analyze the effects of their proposals, it becomes especially important for those involved in the policy process to understand and critique those analyses.

Before we begin, let us state a few terms of clarification. First, "we" refers to the authors of this text. While we draw on many years of training and experience in environmental policy analysis between us, we recognize that we are often providing our opinions, not just an accepted body of thought. We assume that you, the reader, are a policy analyst, and we will refer to you as "you." While we hope that not only policy analysts read this book, using this convention reduces the number of times we have to say "policy analyst." Finally, the client for the analysis (discussed further below) will be "she." Clearly not all clients or decision-makers are female, but choosing one gender reduces the need for the somewhat awkward s/he or him/her.

2. DEFINING THE ANALYSIS

As stated above, we consider a policy analysis to be information designed to contribute to public decision making. "Information," of course, is a very broad term. What kind of information should a policy analysis provide? What information is useful for the policy process? One of the lessons of this book is that there is no one answer to any of these questions. There are many ways that a policy analysis can be conducted, even if the starting point for any of them is the same set of data. Two components are critical to conducting a policy analysis: defining the client, and defining the research question. If these two components are not thought through thoroughly, even the best-conducted analysis may not contribute to the policy process.

2.1 The Role of the Client

A client is the person or organization for whom the analyst conducts a study and to whom the study will be reported. She can be an employer, someone who contracts with an analyst for some work, or an interested party with no financial role. She can be a decision-maker, someone trying to influence a decision-maker, or even another analyst. Why should a client be important to a policy analyst? If information is developed with a particular decision-maker in mind, does that not suggest that the analysis will be slanted to please that decision-maker? If those questions are not coming into your head, they should!

We argue that identifying a client is in fact a critical component of your analysis, for several reasons: first, having a client ensures that your analysis will be used; secondly, it ensures that you are asking a relevant question; and third, it makes clear any potential biases your analyses might have. In essence the client is the audience you are writing for, or the potential user of the information you produce. As always, it is important to keep your audience in mind when choosing the level of analysis to perform and the writing style to use when reporting your findings.

First, if you have a client both involved in the policy process and interested in your analysis, your analysis will get used. Those involved in the policy process are typically eager for information. If the information helps them with their case, they will use it to their advantage. If the information is harmful to their case, it is important for them to be aware of the pitfalls of their stance. If they know of problems with their position, at the least they will develop a better understanding of the obstacles they face. They may even seek to amend their stance or change their negotiating strategy in the face of this information. Having a client for the information (especially a client who pays for it and thus has a financial stake in it) ensures that the information will enter the policy process.

In contrast, if an analysis does not have a client associated with it, it is less certain that the information will be used. First, you will have to gain access to those involved with policy formulation to explain how the analysis might be of interest. Then you have to convince them that your results will be useful to them. If you have a client involved in the analysis from the beginning, you need to do far less convincing from the start.

Secondly, your client can and should help you define the question you will analyze. As will become apparent through the course of this book, an exhaustive (and exhausting) policy analysis could address all of the issues potentially involved in a particular policy. Alternatively, a policy analysis could focus on just one or two key features of a particular policy. For instance, it could look at the overall benefits or costs of a proposal, or it could focus on the effects on a particular segment of society, or it could analyze the physical or biological effects of an action. To the extent that your client knows what issues are most important, or where the policy debate might benefit from greater information, she can help ensure that your analysis will provide information that will contribute directly to the policy debate.

Thirdly, identifying your client is an honest statement of your potential biases. By *bias* we do not only refer to changing the results in order to come up with a desired outcome, but rather affecting the outcome of the study by how the study is designed. The first, in which data may be deliberately used incorrectly, is unethical and scientifically unjustifiable (and frequently unforgivable), and it harms the credibility of any past or future work by the same analyst. There is no valid excuse for this kind of bias. While you will only occasionally be asked/told to slant your analysis, there are lots of subtle ways that a client can "browbeat" or suggest to you that your current results are not acceptable and that you should look for ways to raise/lower the impacts. At this point you will need to make clear that such revision violates professional standards and that such tweaking of the results is likely to call into question the veracity of the entire study. A study that can be proven wrong can hurt the client as well as the analyst.

In contrast to purposefully slanting your results, performing a policy analysis that looks only at a subset of the effects of a given policy is often unavoidable. A thorough and objective, value-free analysis is a worthwhile goal for a policy analyst. Be aware that a selective analysis of the issues can result in the appearance of bias. Further, even when you try to be objective, bias can enter an analysis in subtle, unintentional, and perhaps unavoidable ways.

For instance, a distributional analysis may ignore effects on some affected sectors, because you focused on effects on other groups. Or, a benefit-cost analysis without distributional analysis may imply that effects on individuals or groups are not as important as effects on society as a whole. As another example, if you focus on the biological effects, the study may be accused of ignoring the effects of a policy on people. If you assume that a specific budget has been set aside to protect public health through reduction of exposure to pesticides, the study may be accused of ignoring possible alternative uses of the funds. Choosing a data set to analyze implies that other data sets were not chosen, which implies that you do not consider the other data sets appropriate for your research. Any analysis that focuses on one area is certain to leave out some other areas, and the omission could leave you open to claims of the latter kind bias.

The commissioning or sponsorship of a study by a particular client has implications for the question a policy analyst asks, and since the choice of question inevitably limits the scope of analysis, having a particular client may send signals about the possible slant of the analysis. For instance, an analysis of the effects of clean air regulations by an environmental group is likely to involve different questions than an analysis of the same subject by representatives from coal-fired electrical utilities. If each group of analysts starts off with the same data set and asks the same research question, the groups may well come up with the same answers. The bias results from their starting their work by asking different questions. For instance, the environmental organisations are more likely to be interested in the effects of new rules on human health, while the electrical utilities' organizations are likely to focus on the costs imposed by the new regulations. Both analyses will provide valuable input into the policy process, but they are likely to be used for different ends. Acknowledging the client who sponsored a report signals a recognition that the client influenced the report, even if the influence involved no more than suggesting the topic, providing funding, or giving comments.

At the same time, it is very important not to let the client have undue influence, in order to ensure that you have the freedom to let the numbers come out as they will. As mentioned above, an analysis with distorted or erroneous results can be very harmful both to a cause and to an analyst if discovered, because both will lose credibility. A well documented study with sound methods, good data, and reproducible results will serve both you and the client well, even if the results do not necessarily favor the client.

For the reasons just discussed, having a client for an analysis helps ensure that the analysis will be useful and used. Having a client also aligns you with that client and its policy stance. Being completely unbiased is difficult, if not impossible, but being credible and honest with the analysis is essential for all concerned.

One way to uncover hidden biases and reduce the potential for bias in a policy analysis is to have the study plan reviewed by other analysts. These analysts might include trusted colleagues, or neutral parties at an independent think tank or university. These analysts might even be opponents. It is often better to know potential criticism while you can still do something about it. Such a review of the analysis plan by a wide range of affected publics, and revision to incorporate their concerns, may also be ways of getting these parties to have an implicit stake or even "buy off" on the use of the analysis in the policy process. Further, such a review can also be a safeguard to you in warding off pressure by the client to slant the study design one way or the other.

Box 1-1. Benefits of Including Multiple Stakeholders in a Study

Logging debates in the Pacific Northwest have historically pitted timber and labor interests against environmentalists. In one northeastern California town, though, these groups decided to stop fighting each other. Instead, local merchants, loggers, environmentalists, and county officials formed the Quincy Library Group (QLG) and developed a five-year forest management program for areas within the Lassen, Plumas, and Tahoe National Forests (Quincy Library Group (QLG), 1998). The QLG's collaborative effort resulted in the 1993 Community Stability Proposal, a plan that described how to "promote forest health, ecological integrity, adequate timber supply and local economic stability" (QLG, 1993). The QLG exemplifies how different groups can, through collaborative involvement, use the same model to evaluate different methods and agree on the option that best addresses stakeholder needs.

The early, tense QLG meetings enabled participants to listen to everyone's views: timber interests expressed anxiety about the impact of the industry decline on jobs and the local economy, while environmentalists voiced concerns about clearcutting, preservation of endangered species, and roadless areas. Initially participants felt the only common ground was their belief that the "U.S. Forest Service management failed both the environment and local communities" (Little, 1995), but gradually they established other areas of agreement. This led to the 1993 Community Stability Proposal, which replaced clearcutting with single-tree and group selection logging and supported timber-dependent jobs by directing felled trees to local mills (Little, 1995). In fact, strength of the agreement was confirmed in 1995 when timber mills turned down an opportunity to log in areas the QLG had designated as off-limits (The Economist, 1995). The QLG then lobbied Congress to pass the "Herger-Feinstein, Quincy Library Group Forest Recovery Act," which mandated the Forest Service to implement the Community Stability Proposal (QLG, 1998). The QLG participants recognized the value of collaboration and stakeholder involvement to balance their interests.

2.2 Defining the policy analysis problem

As suggested several times above, there is no one way to do a policy analysis. Any policy issue involves many different facets – how much the policy will cost, what kinds of benefits it produces, who is likely to be affected by the policy (and how they will be affected), what the environmental impacts will be, and so on. If you cannot analyze everything, how do you decide what your subject should be? We suggest two principles that you should use in defining your question: 1. The topic should be negotiated with your client; and 2. The study question should be answerable.

2.2.1 Negotiating with the client

Above, we discussed why a client was helpful for defining the policy question. In choosing your topic, you could start by asking, what kinds of information would be of most use to my client? The answer will depend heavily, of course, on who your client is. Some clients, such as federal agencies required to do environmental impact statements, will need analyses that cover a large range of topics, ranging from ecological impacts to social and economic effects. In other cases, as discussed above, clients will want answers to research questions that are likely support their positions. (Remember, as just discussed, that you can and should do completely unbiased research, even if your research is on a biased question.) While only analyzing a part of the problem could lead to biases, as discussed above, it is not necessarily your responsibility to analyze everything.

As you discuss your study plan with your client, you both need to reach agreement on the specific topics that you will analyze. Some topics may be of more interest to your client, and you will probably prefer to work on others. Neither of you wants to see your time wasted, though. If you conduct an analysis that your client did not expect or does not want, you may face consequences such as losing possible future work, losing your job, or not being paid for the work you did. If you agree in advance on the specific questions that you will seek to answer, and you are able to answer them, you will have done your job.

Of course, if you discover, in the course of doing your work, that there might be reasons to revise your list of research tasks – for instance, you cannot complete a task within the time or budget available, or another, more pressing issue arises – then you and your client might well want to revisit your agreed list of questions. That decision must come from a conversation between you and your client, though: if you have an agreement with your client on what work you will do, you should not change that agreement unilaterally.

You also may have the ability, if you are not directly employed by someone, to come up with an idea that you would like to analyze and to find a client interested in that question. If you can decide who is likely to be interested in the information you are trying to uncover, you can contact those individuals or organizations and see if they might like some sort of relationship with you.

Box 1-2. Same Topic, Different Studies

In 1993 the American Lung Association (ALA) successfully sued the U.S. Environmental Protection Agency (EPA) to require it to review and revise national health standards for ground-level ozone, an air pollutant. The proposal for the new standards came out in November of 1996, and the final standards were issued in July of 1997 after a period of intense debate between environmentalists and industry representatives. The standards, as required by the Clean Air Act, are based only on protecting human health. At the same time, requiring standards that cause economic hardship may be politically untenable. Research can both provide insights into the health and financial aspects of the issue and can be potent political factors in the outcome of the policy process. Two organizations developed and used research on different questions to support their political objectives.

The ALA published several studies dealing with the issue of the effects (in both human health and economic terms) of ozone and particulate matter. One study concluded that ozone is linked to approximately 10,000 to 15,000 hospital admissions for total respiratory illnesses, asthma, pneumonia, and chronic pulmonary disease, and an estimated 30,000 to 50,000 emergency room visits in 13 cities during the summer ozone season (American Lung Association (ALA), 1996). Another study (ALA, 2000) indicated that children are particularly vulnerable to lung damage caused by ozone. A third report (ALA, 1997) concluded that even modest reduction in smog and soot would prevent tens of billion of dollars in health damages each year. All these studies emphasized the adverse effects of ozone on human health.

In contrast, the Reason Foundation (1997), a national research and educational organization that explores and promotes rationality and freedom as the underpinnings of a good society, focused on the costs of EPA's standards. It concluded that EPA's proposed reduction, in mortality would be completely offset by the potential for increases in risk associated with the reduction in annual household income attributed to the high cost of compliance with the proposed standards. EPA would be trading one set of health benefits for another by implementing regulations that do not take economic costs into effect.

These two sets of studies do not necessarily contradict each other. Rather, they emphasize different research questions that support their fundamental missions.

2.2.2 Having questions that can be answered

While having answerable questions may seem like an obvious need, you are almost certain to start discussions over analysis topics with questions that are vague and poorly defined. For instance, how would you answer the question, "What are the impacts to the United States of reducing greenhouse gas emissions?" There are too many undefined parts of that question to ensure a useful answer. Someone needs to specify (a) which impacts are of interest (environmental? economic? public health? ecosystem?), (b) how much greenhouse gases might be reduced, and perhaps even (c) which greenhouse gases might be reduced and (d) by what methods. Questions that are more likely to be answerable might include, "What are the effects on northern boreal forests of achieving the greenhouse gas reductions specified in international negotiations," or "How much are people willing to pay for more fuel-efficient vehicles, and what are the resulting reductions in greenhouse emissions?" Another reasonable researchable question might be, "What are the kinds of impacts likely to result from reductions in greenhouse gas emissions?" For those questions, both you and your client are likely to understand what research results you will deliver.

Your client may or may not want to discuss with you the methods you will use to answer those questions. How much the client wants to know about the details is, of course, a negotiable subject.

As will be further discussed in Chapter 2, one other important aspect of having an answerable question is being able to answer it within the time and budget that you face. Many public policy debates happen in very tight time frames. If your analysis comes out after a decision is made, it cannot influence the decision. You will need to ensure that what you promise to do, can in fact, be done in a timely manner.

3. ETHICAL CONCERNS AND POLICY ANALYSIS

Policy analysis often has a reputation for being objective or value-free. This reputation is not always deserved. As has been discussed, the very choice of which facets of the policy question to be studied can influence the results, just by the choice of what to include and what not to include. For instance, if an analysis looks only at the effects of a policy on one group and ignores effects on others, it can be said that the study suggests that that one group is more worthy of attention than others. You may be asked to draw conclusions from a study in a way that you consider inappropriate. Alternatively, the client may request that you conclude that one policy is unacceptable because it will harm a specific interest group, when you believe that the policy may create enough benefits overall that it is in fact a good thing. What should you do, not only to get out of these situations, but also to avoid getting into them?

These concerns highlight why we have emphasized the importance of the choice of both a client and the question to be analyzed. At those stages – when you are considering whether to work for someone, and when the client and you are agreeing on a specific assignment – you have the ability to accept, to walk away, or to renegotiate. Of course, if you are the employee of the client, it becomes much more difficult to walk away, but you still have the ability to suggest additional analysis issues to the boss and why they are important. Thus you have a dual responsibility, to your client but also to your profession. As will be described in this book, there are professional standards for a competently done policy analysis. These quality standards need to be communicated to an employer when the boss instructs "just give me a number!" Policy analysts don't pull numbers out of the air any more than engineers would. It is also important for you to communicate to the employer or client that an obviously self-serving study will often be dismissed out of hand or, worse, cause a backlash that causes anything the client says to be discounted.

Ethical dilemmas can arise from a variety of sources. These include:

3.1 Whether the client and you are comfortable working with each other

For instance, suppose your client favors increased government regulation in one sector, while you are opposed to government intervention there. Because you two are likely to end up on opposite sides politically, each needs to consider carefully whether you can work together. There can be advantages to this situation, if the analysis provides information that leads opposing parties to agree on a policy. Bipartisan Congressional panels or legislative proposals are sometimes developed for this reason. If they jointly agree that some alternative is in the best interests of the country, even if it causes pain to some specific sectors, the bipartisan nature of the policy limits the ability of one political party to benefit at the expense of another. In other cases, each party must decide whether a possibly difficult relationship is acceptable.

3.2 What policy analysis question is asked

As discussed above (and as will be discussed again), many different questions can be asked about a policy, ranging from the ecological and social impacts to its costs, benefits, feasibility, and legality. Analyzing one of these questions increases the information available about those impacts. If some impacts are better understood than other impacts, the known impacts may receive more attention in the policy-making process than the unknown impacts will. For instance, if the costs of avoiding climate change are better understood than the benefits, then the costs may receive more weight when decisions are made than the benefits will, despite the fact that the benefits may outweigh the costs.

If you are concerned that the way the proposed analysis question is framed will lead to biased or unsound decision-making, you should suggest a broader or rephrased policy question. Frequently there can be a great deal of latitude in the kind of study to be conducted, and you can (and should) help your client understand the implications of the range of answers to the research question being asked.

3.3 How the analysis will be conducted

A policy analysis can be conducted with very little time or money or conducted with enormous resources. You may collect primary data or rely on data from existing sources. How the analysis is conducted can affect the quality of the results and thus the quality of the information provided to policy-makers. For instance, a "quick and dirty" study is likely to be suggestive of the results of a more thorough study and can produce results in a very timely fashion; however, it could be proven incorrect by a more thorough study and therefore might lead to inaccurate information entering the policy process. At the same time, policy-making can suffer from the "paralysis of analysis," using further study of a topic only as a way to delay action. As an analyst, you need to be aware of the limitations of any study that you conduct and to communicate those limitations to the policy-maker. You should balance the usefulness of timely, cost-effective (i.e., cheap) analysis for influencing a policy quickly, compared to the improved quality possible with a longer-term, more thorough study.

3.4 What conclusions will be drawn from a study

Typically, more than one conclusion can be drawn from a study, and the same conclusion can be read in a number of different ways. For instance, studies of the Arctic National Wildlife Refuge (ANWR) in Alaska estimate that between 4.3 and 11.8 billion barrels of oil (BBL) exist in the area, though none is economically feasible to extract if the price of oil is below $15/barrel. At $20/barrel, about 3.2 BBL are considered economically recoverable (U.S. Geological Survey, 1998). Arctic Power, an organization of local groups including the Alaska Oil and Gas Association and the Alaska Chamber of Commerce, argue that this study proves that ANWR is the best remaining exploratory site for oil in North

America (Arctic Power, 2000). In contrast, the Natural Resources Defense Council (an environmental organization) divided the 3.2 BBL by current consumption and conclude that ANWR will provide about 6 months' worth of energy to the U.S. (Natural Resources Defense Council, 2000). Is a supply of 6 months' worth of oil critical for protecting the U.S. against possible oil shocks, or is it a mere drop in the bucket of the U.S.'s insatiable appetite for oil? In this case, relatively objective information can be (and has been) interpreted in very different ways. In other cases, conclusions are drawn from some, but not all, of the results of the analysis, such as when the impacts of a policy on one group are emphasized over the impacts on another.

You may want to discuss with your client, in advance of undertaking the study, whether the study will remain solely in her control, or whether you have the ability to make the results more widely known. If the full study is made public, then the full set of conclusions from the analysis can frequently be drawn as others read and provide critique of the study. If your client has control over the study, it is somewhat more likely that the results will be used selectively; at the same time, it means that she may be more likely to accept analysis with possibly unfavorable results, since she will not necessarily have to reveal them.

Before the study is undertaken, you should discuss with your client how the results of the analysis will be distributed and who will be responsible for them. Will you be able to discuss or publish the results of the study without her authorization? Who will be required to answer questions about the study? Having those questions addressed in advance will help prevent conflict after the study is completed.

3.5 What recommendations will be made from the study

Just as different conclusions can be drawn from the same data analysis, different recommendations for action can arise from different (or even the same) conclusions. Should you provide recommendations to your client, or should she be allowed to decide on actions by herself?

Of course the immediate answer is, "it depends." On some occasions a client has a position and wants supporting information, not reasons to change the position. On other occasions the client is looking for ideas on what position to take, or how to modify a proposal, and may welcome suggestions. Whether recommendations need to be made is a matter to be worked out between you and your client.

If you are to make recommendations, the next question becomes, on what basis should the recommendations be made? Your client may well have specific objectives that she wants to achieve; one possibility, of course, is to make recommendations that fulfill those objectives. If you disagree with the objectives of your client, though, making recommendations to fulfill her objectives could become an ethical dilemma. For instance, your client may be pursuing a subsidy for a specific interest group, and you may not support a subsidy for that group. You may still be comfortable in performing a study for the client: if both of you can agree on an analysis question to conduct (for instance, the effects of a policy on her interest group), the analysis may not in itself pose a problem. Still, you may not want to make a recommendation from the study that you client would like included.

Here, you may want to suggest that the recommendations be left out of the study itself, while your client can be permitted to draw conclusions or make recommendations separately from the study, through press releases or other means. Additionally, as before, a discussion of expectations between you and your client before the study is undertaken may help avoid future problems.

As an analyst, you have to decide when you are comfortable with an assignment, and how to handle assignments with which you are not comfortable. The decision has to be individual. Probably the strongest message that can arise from this discussion is to think ahead. Where might problems arise? Can you and your client come to agreement about how to handle possible problems? You must search your own conscience to decide whether to pursue assignments that may violate some aspect of your beliefs, and we hope that your client will respect those beliefs.

A key element to keep in mind is the distinction in many agencies between the analyst and the decision-maker. Neither the analysis nor the analyst will make the decision. The analysis is information, and the analyst an employee. The ultimate decision-maker is often the agency head or an elected public official whose responsibility is to make such decisions. Equally important is for the decision-maker not to meddle in the technical details of the study. The results are the results, and the decision-maker is free to use or ignore them. This distinction is often referred to as one of line (decision-maker) and staff. While the distinction is sometimes blurred, keeping this difference in mind can spare both parties in the policy process much grief and embarrassment.

4. SOME FEDERAL REQUIREMENTS FOR POLICY ANALYSIS

Not only is policy analysis usually a good idea, in many cases it's the law. The U.S. government is required to conduct analysis of the impacts of many of its activities. The intention is usually to promote more informed decision processes. Some examples of these requirements are:

4.1 The Electric Consumers Protection Act of 1986

This act amends the Federal Power Act by changing the requirements for licensing or relicensing of hydropower facilities. The "equal consideration" clause mandates that the Federal Energy Regulatory Commission (FERC) give equal consideration in its licensing decisions to the value of non-power benefits of a river as well as the benefits derived from power.

4.2 Executive Order 12291 (Feb. 17, 1981)

This Executive Order is intended to streamline the regulatory process by instituting a procedure for evaluating the creation, impacts, and effectiveness of present and future federal regulations. One of the requirements, for all agencies to the extent permitted by law, is a benefit/cost analysis of regulatory priorities, objectives, actions, and different alternatives.

4.3 Executive Order 12898 (Feb. 11, 1994) – Federal Actions to Address Environmental Justice in Minority Populations and Low-Income Populations

This Order requires that all agencies strive to achieve environmental justice by conducting an analysis to determine and address "disproportionately high and adverse human health or environmental effects of its programs, policies, and activities on minority and low-income populations in the US and it's territories and possessions."

4.4 The National Environmental Policy Act of 1969

NEPA's major contribution to the world of policy analysis is the requirement that all Federal actions that "significantly affect the quality of the human environment" must be examined in an Environmental Impact Statement (EIS). The EIS should address: impacts of the proposed actions; unavoidable adverse effects caused by implementation; alternatives; the relationship between short term uses and long term productivity; and any irreversible commitments of resources resulting from implementation of a proposed action. In addition, many states have requirements for policy analysis for environmental activities, including state-level equivalents to the National Environmental Policy Act.

4.5 U.S. Water Resources Council—Principles and Standards for Water Resource and Related Land Use Planning (1973)

The Principles and Standards established a detailed planning process, including formulation of alternative plans and an analysis of trade-offs between the alternatives. For each alternative project the agency has to conduct a benefit-cost analysis for national economic development objectives and for environmental quality objectives. The effects on other factors like regional development and social well being should also be indicated, but a detailed analysis of them is not required.

5. LESSONS FROM THIS CHAPTER

This book is intended to help a policy analyst to do the job, as well as to help those who read policy analyses to understand what they mean. If a policy analysis is to be information provided for the policy process, then it needs to be relevant to a current issue, and it needs to have an avenue into the policy process. The first of these is rooted in the discussion of defining the analysis question. A policy analysis can take many forms, and finding the question in most need of being answered helps ensure that your results will be useful. Finding an avenue into the policy process for the study is greatly facilitated by a client. The client is likely to be involved in the issue and thus recognize the most important analysis questions, and the client is likely to make use of the results.

Because policy analysis can influence formulation of laws, regulations, and public projects, and therefore the lives of millions of people (as well as the environment), ethical issues and questions of personal values will be almost unavoidable. Even if an analysis is done with the most careful, objective, unbiased methods possible, and is written up so that all results and conclusions are absolutely unassailable, moral questions can still arise about the choice of question analyzed or the role of the client for the analysis. Additionally, policy analysis is often done with a limited budget and a very short time frame, either of which by itself can lead to limitations in the analysis; how those limitations are addressed can lead to problems. There is no way to avoid these dilemmas and decisions. Still, it is our hope that making the fledgling analyst aware of these concerns will give you the opportunity to consider in advance different ways of addressing them.

Chapter 2

Scoping the Policy Analysis

As will be discussed in Chapter 3, one of the key roles of a policy analyst is to provide information on a wide range of impacts of a policy alternative. What are the biological and social impacts of a policy? How much is it likely to cost, and do the benefits of the activity outweigh the costs? Future chapters will describe some of the tools that you can use to provide those pieces of information to a decision-maker. This chapter will first discuss a number of issues relating to defining the scope of your analysis. Properly determining the scope of the work – how much detail is required, from where the data will come, what the time frame or the length of the analysis will be – will simplify and focus your analysis. Therefore proper scoping will increase the chance your analysis will satisfy and be used by the client.

Many of the pieces of information you would like to have, if they are to be completely accurate, would require years of research by highly trained technical experts, and even then the experts might not reach a conclusion on the effects they are studying. Typically, a policy analyst cannot conduct the detailed studies truly necessary to understand the effects of a policy. Either time, resources, or technical skills may not be available to you. However, poor or inappropriate information can lead to bad policy decisions. You would like to provide the decision-maker with the best possible insights into the effects of each alternative, especially if the decision on which alternative will be chosen hinges on these effects. As a policy analyst, you will need to balance the need for speed with the necessity of accuracy.

This chapter will highlight some of the issues you will need to address before you begin your analysis in detail. The key message is that you *must* fit your analysis to the time and resources available to you. If your client needs extremely precise results, and those results are not available from existing research, then you must inform the client of the resources necessary to get the degree of precision she requests. If your client will be satisfied by knowing what the effects are likely to be, without needing to know the magnitudes, you are much more likely to be able to complete your work quickly and easily. You will do neither yourself nor your client any favors by promising a highly sophisticated analysis that you cannot deliver in the time frame requested. Either you will be sleepless and underpaid, or your client will be dissatisfied with your work. It is much less likely that your client will be upset if you provide specific, detailed results when general comments would have served, except that you spent time and resources she might have preferred that you use elsewhere.

1. AGREEING UPON AN ANALYSIS BUDGET

There are several issues that will need to be resolved when you first discuss the analysis with your client. Will you spend six months or six hours on the study? Will your budget enable you to hire fellow analysts, or will you (and you alone) have to do the work? Will you be able to develop new data sources, or must you rely on someone else's study of an issue that's sort of (but not quite) similar to your issue? If you have to submit a bid to conduct a study, as consulting firms frequently do, you will have to develop an explicit budget and time line for your work, as well as a description of your expected final product. If you work directly for your client, your time and budget constraints may be negotiated directly with your employer. Regardless, these issues must be resolved before the work begins, to ensure that your time and effort are spent as your client wants.

Often the scope of the project will be determined even before you know that you will work on it: your client may want a result immediately, or she may have already decided that a multi-year, multi-million-dollar analysis will be necessary. In other cases, you will have an opportunity to suggest the appropriate scope for the analysis. Money spent on policy analysis, from a societal perspective, is money not spent on other activities, such as implementing policies. The purpose of the analysis is to improve the decision that will ultimately be made about the policy. At a theoretical level, a study is worth conducting if the benefits of the study – the improvement in the decision – outweigh the costs -- the time and money invested in the analysis. Under some circumstances, conducting a study is unlikely to improve a decision. Several factors should influence both whether to conduct an analysis and what the scope of the analysis should be:

1.1 Might a decision change because of your analysis?

Is the decision-maker looking for the best alternative? It is possible that the decision is already determined – perhaps because a law mandates an action or because the decision-maker has already decided. If so, then nothing will change because of the study, and you might have more influence by working on another topic. On the other hand, an analysis may nevertheless be useful if future actions might change or mitigate the effects of the policy.

1.2 How well understood at this time are the impacts of the policy?

If the effects of the policy are completely unknown, then there is value in determining whether those effects are likely to be large or negative. If the policy is well studied, then another study may not be the best use of society's money. Sometimes, though, policy-makers may not respond to a study because they may not be convinced that it is really correct. Having studies that confirm the original results could add important credibility to the original analysis.

1.3 How large are the impacts of the decision relative to the amount of money necessary to conduct the study?

If the differences in impacts across alternatives are large – if millions of dollars, acres, or lives are at stake, for instance – then spending at least a small amount (and possibly a large amount) studying impacts is likely to be worthwhile. If the differences in impacts are small, though, then spending a lot of money on a study is not socially productive.

Those factors are useful for you and your client if you are deciding the socially optimal level of analysis to conduct. Is it likely, though, that the socially optimal level of analysis will be the basis of either your or your client's decision on analysis scope? In reality, you and your client will each have your own objectives in this negotiation. At an individual level, she may want a highly credible and completely thorough analysis done in almost no time for almost no money, while you may want to be well paid for a long period of time and not have to work very hard. Both of you are likely to recognize the unfeasibility of these positions. Settling the scope of the analysis will require negotiation and careful communication to make sure that you both understand the work to be done and the time and budget you will face.

Analysis can also be conducted for political reasons. A policy-maker may be reluctant to make a decision because she is uncertain or indecisive, or because she does not approve of the proposed alternative, or because she hopes that another study will overturn results of the last study. In these cases, claiming a need for further study can bring policy implementation to a halt at least until the study is completed (a phenomenon sometimes called "paralysis by analysis"). While you may not approve of the decision to postpone, keep in mind that you are not the decision-maker. It is her job to face the consequences of this decision. The analyst does not get blamed for doing good analysis, though the decision-maker can get blamed for requiring more analysis than a policy needs.The remainder of this chapter will discuss some of the issues you will run into as you determine the scope of your analysis. This list is suggestive rather than conclusive: other issues may arise as you become more experienced in policy analysis. In all cases, we encourage you to discuss in advance with your client how to balance the tradeoffs between limited time, money, and resources, and the desire for the best possible information.

2. USE OF THE "WITH AND WITHOUT" PRINCIPLE

One of the fundamental principles of policy analysis is to separate out the effects that are due specifically to the policy action from those that will occur without the policy action. In essence, we wish to know increases in the level of environmental quality that would result from implementing a particular regulation beyond the levels that would have occurred even if we do nothing. For example, if we are trying to estimate the benefits of requiring a curbside recycling program, it is only the additional recycling induced by this program that is a benefit, not the total number of amount of recycling after the policy is implemented. Some amount of recycling is already occurring via store drop-off points and recycling centers. If the city already has a "pay as you throw" policy of pricing garbage collection based on volume set out at the curb, then over time, households may take actions to increase the amount of recycling, even in absence of the curbside recycling program.

Figure 2-1 illustrates this point. Under the current conditions, in year one, households recycle 1,000 tons. In ten years, that recycling is expected to go up to 1,200 tons, even if no curbside recycling policy is required, because of increased environmental ethics and the desire to reduce their trash bill. However, if the curbside recycling program is mandated, recycling will become more convenient and the amount will rise to 1,500 tons in ten years.

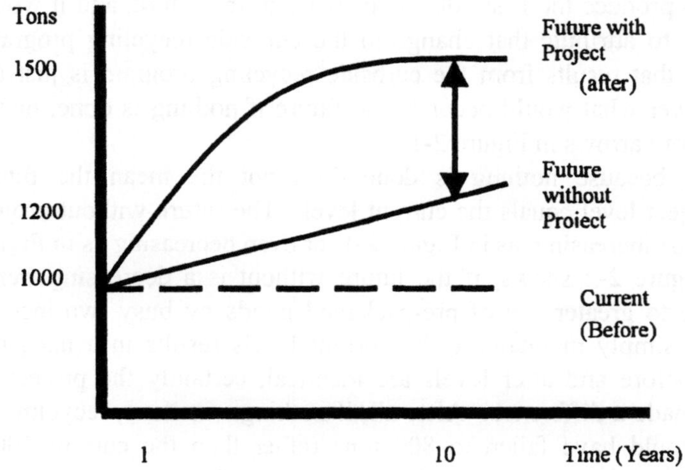

Figure 2-1. Correct Comparisons Using With versus Without

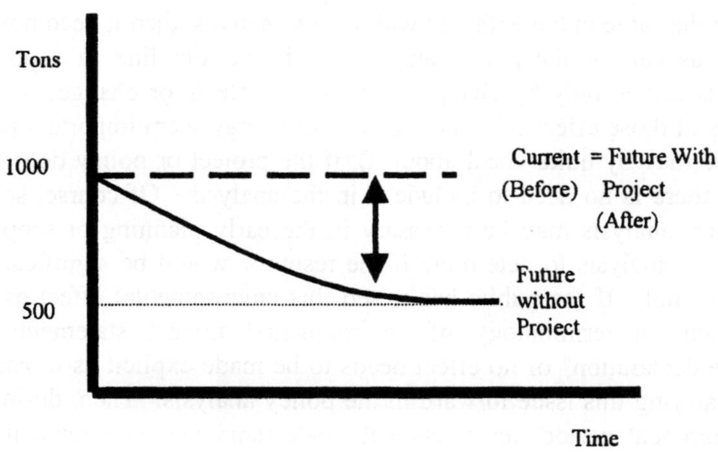

Figure 2-2. Correct Comparisons Using With versus Without with No Change Baseline

Analysts often become confused over whether the benefits of the program should be the difference of the before program minus the after program (500 tons) or the difference between the with and without program (300 tons).

Clearly, it would be erroneous to compare before and after, since part of the gain in recycling (200 tons) would have occurred even if the curbside recycling program had not been implemented. No new program is needed to produce the first 200 ton increase in the future, and it would be incorrect to attribute that change to the curbside recycling program. The change that results from the curbside recycling program is just the increment over what would occur in the future if nothing is done, or the heavy two way arrows in Figure 2-1.

But just because nothing is done does not the mean the future without-project level equals the current level. The future without project trend could be increasing, as in Figure 2-1, or even decreasing as in Figure 2-2. As Figure 2-2 shows, if the future without is a decreasing trend, perhaps due to greater use of pre-packaged goods by busy two-income households, simply maintaining the current levels results in a net gain. While the before and after levels are identical, certainly the project or policy has made a difference. After all, if nothing was done, recycling in ten years would have fallen to 800 tons rather than the current 1000. Thus, avoiding a future loss is a valid program benefit.

The "with-without" principle is also very helpful in zeroing in on exactly what resources need to be fully evaluated in policy analysis, and which can be ignored. If the level of a particular environmental effect would be the same in the with and without alternatives, then it need not be analyzed as part of the policy analysis. The bottom line in a policy analysis is driven only by changes in types of effects or changes in the quantities of those effects. While a given effect may seem important (and the public may by quite vocal about it), if the project or policy does not affect it, there is no need to include it in the analysis. Of course, some preliminary analysis may be necessary in the early planning or scoping phase of an analysis to determine if the resource would be significantly affected or not. If the public has raised this environmental effect as an issue, then, in terminology of environmental impact statements, a "negative declaration" of no effect needs to be made explicit as a reason for not carrying this issue forward in the policy analysis. Then, during a public comment period, reviewers will understand that you recognized this issue, but that it is irrelevant to the choice of policy alternative because it does not vary across the alternatives being considered.

Box 2-1. With or Without NAFTA? That Should Be the Question

The primary objectives of the North American Free Trade Agreement (NAFTA) are to increase trade and facilitate the flow of goods and services between Mexico, the United States and Canada (American Chamber of Commerce of Mexico, 1997). Some environmental organizations, such as Sierra Club and Public Citizen, have expressed concern about levels of regulatory enforcement in the Mexico-United States border region and have asserted that NAFTA has caused "environmentally destructive and socially inequitable investments" (Sierra Club, 1996). NAFTA supporters, on the other hand, argue that the Agreement has encouraged Mexico to implement substantive laws mandating environmentally responsible actions. Although NAFTA supporters and opponents assess the Agreement's impacts, or the "with NAFTA" scenario, neither side supports its claim with a baseline reference, or the "without NAFTA" scenario. Only with a baseline reference -- an estimate of the border region's environmental quality in the absence of NAFTA -- can the environmental effects of NAFTA be determined.

Few studies have evaluated the condition of the environment in the absence of NAFTA. In fact, the U.S. General Accounting Office (Hecker, 1997) testified before Congress that evaluating NAFTA's true impacts is challenging because it is "hard to isolate the impact of the agreement from other trends and events." Analysts at the Institute for International Economics, a research institution, recently evaluated NAFTA's environmental provisions, conditions at the border region, and trends in North American environmental policy. They concluded that environmental problems have not resulted from NAFTA (Hufbauer et al., 2000).

A lack of consistent pollution monitoring data has also impeded efforts to establish a baseline reference for the border region. The North American Commission for Environmental Cooperation (CEC), a tri-national organization created by a NAFTA environmental side accord, has compiled annual pollutant release and transfer registers based on data from 1994 to the present in Canada and the United States. The CEC has thus far been unable to include Mexico in the registers because the Mexican reporting system did not collect data prior to 1997. In fact, the CEC reports that most data from the first reporting year, 1997, were inaccurate, and training and outreach will be conducted to improve Mexico's inventory data (North American Commission for Environmental Cooperation, 2000). Additionally, comparing emissions data from the three countries poses a challenge because of differing reference points, such as program starting dates.

The NAFTA environmental debates demonstrate the need and the challenges associated with developing the "with—without" principle and emphasize the value of baseline references for assessing regulatory impacts.

3. HOW PRECISE SHOULD THE RESULTS BE?

By its nature, a policy analysis seeks to predict the impacts of a policy. With enough effort the prediction can be given with great precision, but with less effort the expected direction of a change may be provided. For instance, a policy analysis may conclude that an increase in the gasoline tax, by reducing gasoline consumption, will improve air quality. In some cases, knowing only the direction of the effect of the gasoline tax on air pollution is sufficient for the purposes of the analysis – for instance, if you are asked to determine if there are any adverse environmental impacts of a gasoline tax.

In other cases, the magnitude of the effect is necessary. For instance, suppose that you are analyzing alternative ways to achieve a specified reduction in air pollution in a city, and you are trying to determine how high a gas tax needs to be to achieve that standard. Here, you must determine how much less gasoline people will buy in response to the tax, and how much pollution is avoided due to that reduction. (You might also need to determine the pollution caused by people switching to buses or alternative-fuel vehicles, if you expect those effects to be large.)

Each of these analyses serves a slightly different purpose and could be perfectly appropriate for your client, depending on her requirements from you. They are likely to involve different time requirements and different research methods. For the first case of a qualitative analysis, a conceptual or theoretical model of the relationship between a gasoline tax and air pollution may permit you to identify the kinds of effects that would lead a gasoline tax to reduce pollution (or, more importantly, what effects might lead that tax to increase pollution). Alternatively, you might compare emissions from cars in areas with varying gasoline prices, to estimate whether emissions from cars are lower where gasoline prices are higher. (You may end up with an actual estimate of the relationship, or you may only have a survey with rankings of the price and rankings of emissions.) In the second case, you will need several kinds of data: first, on the relationship between gasoline price and quantity purchased (the demand elasticity or the slope of the demand curve); secondly, on the change in pollution due to the change in gasoline consumption; third, on possible substitutes that people will use instead of gasoline (i.e., the elasticity of substitution); and fourth, the change in pollution due to the use of those substitutes.

Frequently the precision of the results you can provide will depend on your access to time and money. For instance, Loomis (1993) includes analyses of land planning alternatives conducted by the U.S. Forest Service (USFS) and the U.S. Bureau of Land Management (BLM). BLM historically has received far less funding than the USFS, although it manages about 50% more land than the USFS. As a result, when BLM determines the effects of an alternative on recreation, it uses an estimate of the number of acres of land available for different kinds of recreation (such as primitive or developed). In contrast, the USFS is typically able not only to calculate the acreage available, but also to estimate the predicted use (and the value of that use). Because budget limitations have much more of an effect on BLM than on the USFS, its analysis of alternatives is much more sketchy.

Even if you can provide no more than a direction to an effect – for instance, determining that air pollution will go down from a gasoline tax – you must be very careful how you derive that conclusion. Unintended consequences can lurk in many places, and you must identify them, to the best of your ability, if they appear to be likely. For instance, it seems highly likely that a gasoline tax, by reducing gasoline consumption, would reduce automobile emissions. Suppose, though, that, in response to the gasoline tax, people purchased and drove dirtier cars run by other fuels (such as diesel). Alternatively, perhaps more worked at home, used more electricity through their home offices, and increased pollution through electricity generation. If these effects are small, they need not be of concern. If the effects are large, though, that increase in the gasoline tax could possibly lead to increased pollution.

Unintended consequences can arise in various ways. As noted above, they can arise due to unpredicted effects, such as when a policy induces an increase in the use of less desirable goods. More generally, they can arise whenever a policy measure is attached to something indirectly related to the problem, rather than to the problem itself (such as taxing gasoline rather than pollution from automobiles, when the problem is the pollution, not the gasoline *per se*). Additionally, even if a policy targets the problem directly, if there are unintended effects in other parts of society, society might be made worse off. This "Theory of the Second Best" (Lipsey and Lancaster, 1957) states that, when there are multiple imperfections in the economy, correcting only one of them may make society worse off than if no action were taken. For instance, taxing the pollution from cars, but not taxing the same pollutants if they came from electrical utilities, might stimulate people to drive electric cars. If the pollution from electricity generation is less than that from gasoline use, then pollution will be reduced, though cars will likely be more expensive. If, however, pollution actually increases due to electricity generation, and cars become more expensive, then society may not be better off from the tax on auto pollution.

Having imprecise results is not necessarily a flaw in a policy analysis, as long as both you and your client understand the limitations of your analysis. (It is also frequently true that results that appear precise do not deserve that impression. For instance, if your answer to a problem is 5, plus or minus 200, then saying that the answer is 5 does not indicate the uncertainty surrounding your answer.) Time, data, and resource limitations may not permit you to develop detailed results. Understanding the direction of effects, or even rough estimates of the magnitudes of the effects, can still improve the decisions of a policy-maker: you can identify where impacts are likely to be large, and where they are likely to be small enough not to be of concern; you can highlight possible sources of unintended consequences; and you can flag places where more detailed analysis might be especially helpful. Of course, if you are given the opportunity and resources to conduct a more in-depth study, you can provide more insights. If time is limited, though, having some results is better than promising great results when the policy is no longer of interest.

4. ORIGINAL VS. SECONDARY DATA SOURCES

Time and resource limitations will also affect your ability to assemble the appropriate research materials. Will you have sufficient time and library access to perform a full review of the existing literature, or will you only be able to examine the materials you happen to have at hand? Will you collect original data, or will you draw on data from existing studies?

To provide the information your client will need, your background materials and data sources will take many forms and will come from a variety of sources. All data have their limitations. One of the major skills you should develop as an analyst is determining when you are using data or existing findings in legitimate ways, and when you are using them in ways for which they were not intended and are not suited. This rule applies both to data that you use from other sources and to data that you collect yourself: in either case, the data may not fit your purposes exactly, and you will need to determine whether imperfect data are better or worse than no data.

Collecting original data sounds like it would be superior to using other people's data, but that superiority depends, as in so much in policy analysis, on the time, expertise, and resources available to you. Collecting data can be very expensive and time-consuming, and getting worthwhile results requires careful planning on how to gather the information. If your resources are limited, you are likely to be better off using existing studies and data as much as possible. If, however, you cannot identify any prior studies of your subject of interest, you have a choice of either obtaining the necessary resources to conduct the study on your own, or you can try to extrapolate from existing research that is not directly related to your study.

For example, suppose you are studying the effects of timber harvesting on elk herds in Idaho, but the only existing research you can find is on the effects of timber harvesting on deer in Wisconsin. You could commission a team of biologists to study the effects of timber harvesting on elk in Idaho; that study might take one or more years. Alternatively, you may try to extrapolate some useful information from the Wisconsin deer study. In that case, you need to determine whether deer and elk are likely to respond the same way to timber harvesting, and whether timber harvesting in Idaho is likely to have similar effects on wildlife as timber harvesting in Wisconsin. Are there scientific reasons that the deer data might provide an upper or a lower bound on effects on elk, or that Wisconsin data are likely to be higher or lower than Idaho data? Having some ability to compare between data sources may provide you with information and understanding that you previously did not have.

The same problem can arise with data that you generate yourself if your time or budget do not permit you to gather the "right" data. For instance, suppose that you would like to survey people on their attitude toward protecting air quality over Voyageurs National Park in Minnesota, but you only have the time and resources to ask people in your city. Will your sample of local people give you answers representative of those of people in other regions? Are you better off working with this sample, or extrapolating from existing research?

There are no general answers to these issues; the answer will depend on the level of existing research, on your ability to conduct a study, and, again, the need to get the most precise answer possible. You should discuss carefully with your client the advantages and disadvantages of existing work compared to new data sources, and the limitations resulting from your study however you develop it.

Box 2-2. Benefit Transfer and Passive Use Values of Wild Salmon

Avoiding extinction of endangered species is a source of existence or passive use value, defined as the benefit from knowing a resource exists even if no use is made of it. People who do not fish, participate in river recreation, or eat salmon may still pay something for a river with native or wild salmon. Passive use values are public goods, since these benefits can be simultaneously enjoyed by millions of people across the country (Loomis, 1996a). Wild stocks of Snake River Salmon have passive use value.

Passive use values can be estimated using the contingent valuation method (CVM; see Chapter 9). Unfortunately, CVM surveys take months to do and can cost up to $500,000. How can an analyst address this topic without doing an original study? Benefit transfer, "...the transfer of existing estimates of nonmarket values to a new study [site] which is different from the study for which the values were originally estimated" (Boyle and Bergstrom, 1992:657), was developed for just such purposes.

There are three approaches to benefit transfer: (a) use one estimate from a study that closely matches the application; (b) take an average of point estimates from related studies; or (c) transfer a benefit function such as a demand curve or willingness to pay equation (Loomis, 1992).

Loomis (1999) used these three approaches to transfer benefits from the existing literature to estimate the change in existence value from removing dams for restoring salmon populations in the Lower Snake River. While none is perfect, they suggest a likely range of values for restoring salmon.

One approach, using (a) above, matched the change in salmon populations due to dam removal to an existing study that valued a similar size change in salmon (Loomis, 1996b). The resulting lower estimate of the gain in passive use value would be about $205 million per year to non-user households in the Pacific Northwest and California. Another approach, using (b), took a simple average of estimates from the literature. The result estimates the net gain of dam removal as $261 million per year.

The (c) approach estimated a willingness to pay (WTP) function for salmon from four studies of West Coast residents' WTP for increasing salmon populations. From these studies, dam removal is estimated to produce $1,277 million in annual passive use values.

Two of the three estimates are quite close ($205 and $261 million), while the third approach results in a much larger estimate. Nonetheless, the three approaches allow the analyst to bracket the likely magnitude of passive use values for salmon in the Snake River without doing an original survey. Since passive use values for increasing endangered salmon in the Snake River are certainly greater than zero, the benefit transfer approach avoids omission of this category of benefits. Omission is often taken as an implied zero value.

5. CONCLUSION

The most important part of scoping your analysis is determining the consequences of being wrong. If, for instance, your topic is one that will not have major effects on anyone, or if your client is not likely to make much use of your work, then not making an exact prediction of the effects of that policy will not matter very much. On the other hand, if millions of dollars or many lives may rest on the results of your study, then you will want to ensure the greatest possible accuracy. In either case, your personal and professional optimization problem is to come up with the best possible analysis that you can for your client, given the time and budget constraints imposed on you. If you cannot produce the level of precision in the results that your client wants when faced with the constraints that you face, you should discuss with your client whether you should even continue with your study. The question, in general, is not whether imperfect results are better than no results at all; all results are imperfect, to some degree of accuracy. The question instead is whether the level of precision attainable will make the study worthwhile for your client. As will be discussed in Chapter 7, a technique called "sensitivity analysis" can aid in uncovering which of the many imperfectly measured variables might have a substantial influence on the ranking of alternatives. It is that variable that would most deserve more analysis attention or greater data collection, not all imperfectly measured variables. Thus the key in scoping is to tailor the level of analysis effort to the magnitude of the values at risk in the decision. A productive analyst's motto might be something like, "It does not have to be perfect, but it does have to be done."

Chapter 3

Decision Criteria and Decision Methods for Policy Analysis

Suppose that you were a decision-maker who wanted to do what was best for society for an environmental policy controversy. Perhaps it relates to the siting of a polluting facility that promises to bring with it job opportunities for the poorest people in your community. Maybe you are determining how to protect an endangered species. Or perhaps you are negotiating reductions in greenhouse gas production to reduce the effects of global climate change. How should you make your decision? What kinds of information would you like to have before you decide? In other words, what information would you like your policy analysis to produce?

First of all, note that there are two different questions that a decision-maker is likely to face: what kinds of information are useful, and on what basis a decision should be made. The first we will refer to as *decision criteria*: the factors that go into a decision. The second we will refer to as a *decision rule* or *decision method*. The criteria could be many and varied. The decision rule or method assembles the decision criteria into a bottom line – the decision.

It is important to remember that the decision-making authority ultimately rests with your client, and you are the analyst/advisor. Your primary responsibility as an analyst is to assemble the data, make predictions of the effects of each alternative on each of the decision *criteria*, and then make them available to your client in a usable form. Another role that you may provide is to apply a decision *method* supplied by your client. You may also present the client with the range of decision methods that could be applied to this problem. Typically, you will take the criteria you have developed, run them through the decision method algorithm chosen by your client, and give the results to your client. At this point, your client can do what she wants with the information.

1. TO WHAT ARE THE CRITERIA APPLIED?

Before we discuss decision criteria, we need to take a step backward in the process and discuss the above question: on what basis are we looking for impacts? A critical element of policy analysis is the formulation of alternatives. Rarely does a decision-maker want only one choice; after all, if there is only one choice, there is not much deciding to do. Put another way, if you only provide one alternative, you have made the decision for your employer. Indeed, there is usually a wide spectrum of alternatives available to your client, either in terms of the end result or in terms of the means to reach the end result. In some cases, the alternatives are provided to you by your client; in other cases, one of your jobs will be to identify alternatives. In the first situation, you can go straight to conducting your analysis; in the second case, though, you will have to give careful thought to the range of options you might consider. This issue is discussed further in Chapter 4.

One alternative that should always be included is the "no action" alternative: what does or will the world look like if nothing is done? Then, alternatives are likely to include the range of proposed policies under consideration. Including a wide range of alternatives will increase the decision-maker's understanding of the problem, enable her to demonstrate to various interested parties that their concerns were studied seriously, and increase her ability to find a good outcome. Thus, if you are not given a wide range of alternatives, you should suggest expanding the set to her.

We will remind you here that your analysis of your alternatives – that is, the development of your estimates of the decision criteria – needs to be based, not on a comparison to what the world is like now, but on a comparison to what the world would be like without the alternative.

Now that you know what you will be analyzing – some set of alternatives – the next question is, what will you analyze? The *decision criteria* that follow provide a starting point for ideas on what you need to include.

2. DECISION CRITERIA

What factors are important to understand for environmental policy questions? Clearly the answer to this question will depend on the specific policy being analyzed and its associated laws. However, a complete analysis frequently appropriate for many environmental issues includes:

2.1 The physical/biological/scientific/technical effects of an action

If a policy is undertaken, what will the impacts be? Will habitat be improved or harmed? Will water quality improve or get worse? Will traffic congestion get better or worse? Prediction of the effects of an action is critical information for making a knowledgeable decision about whether to undertake the action. This information on physical and technical effects is usually the foundation for evaluating the next feature. Thus, this step must be performed prior to initiating the next step, not in parallel with step 2. (Attempts to answer the next issue without addressing the first issue have been the downfall of many policy analyses.)

2.2 Economic efficiency (benefits and costs)

Here we refer not only to goods for which market prices exist, but also to other non-marketed goods/services that influence the overall level of social well being. Will the policy overall have a positive effect on society, or will the costs of the policy outweigh the benefits? This item is typically analyzed using benefit-cost analysis, the topic of Chapter 7.

2.3 Distributional equity

Public policies have disparate effects on different people, whether by location, ethnicity, income, or occupation. Who gains and who loses from this action? The distributional consequences of a policy typically indicate a great deal about who supports and who opposes that policy.

2.4 Social/cultural acceptability

Will some people be offended by a particular action? For instance, is a proposed road going to disrupt a site sacred to Native Americans? Is hunting wild horses and burros acceptable for control of their effects on rangeland?

Is it acceptable to put a new chemical plant in a predominantly African-American community that already bears a disproportionate share of pollution? A policy-maker will want to understand why a policy may be opposed, including opposition due to violation of social norms.

2.5 Operational practicality/administrative feasibility

Can an activity be moved from theory to reality? For instance, will marketable permits work for reduction of greenhouse gas emissions if emissions of greenhouse gases cannot be monitored accurately? Will an agency have sufficient personnel and budget to enforce a rule prohibiting the taking of endangered plants from private lands? If a policy cannot be implemented effectively, will there be adverse unintended consequences?

2.6 Legality

Are the alternatives possible without changing existing laws? Even if an idea sounds wonderful, if it would require changing the state's constitution, existing laws, or existing regulations, it may be difficult or impossible to implement without great expenditures or a lengthy time frame. Thus, while a policy alternative can certainly identify laws that should be changed to improve social well being, this is usually not the primary result of a policy analysis. Most practical policy analyses have to find solutions within existing laws and regulations.

2.7 Uncertainty

Are the effects of a policy certain to happen? If there is ambiguity about the effects of the policy, is the range of uncertainty acceptable, or does the range of possible outcomes change the results of the analysis dramatically? That is, does the ranking of alternative solutions completely switch with plausible ranges of critical variables? If uncertainty is prevalent in the area of study, it will be important to present ranges of results as well as just the mean. If there is any risk aversion, then a decision-maker is probably not indifferent between two alternatives with the same average effects but different ranges of outcomes. For instance, suppose that two alternatives have a mean or expected value of $50 million in costs, but one's possible costs range from $1 million to $99 million, and the other has possible costs ranging from $48 million to $52 million. The second alternative clearly provides more certainty and thus might be more desirable. As mentioned before, the climate change debate is frequently framed as having great scientific uncertainty: the effects could be big, or they could be non-existent. If they are big, then we might well want to change our current behavior in dramatic and expensive ways; if they are non-existent, society would probably not bother to impose those costs on ourselves, since there would be no associated benefit. Because the range of possible outcomes implies radically different policies, reducing the degree of uncertainty could lead to more consensus on what actions to take. Thus, if the preferred alternative is highly sensitive to variables whose uncertainty can be reduced by greater data collection or more detailed analysis, it may be worthwhile to undertake the additional data collection or analysis.

This list is meant to be suggestive rather than conclusive. Some of these topics may not be relevant for some issues, and your particular analysis (as determined in coordination with your client) may include only a small selection of these categories. For the ultimate decision-maker on an environmental policy, it is likely that virtually all these factors will come into play.

Box 3-1. Uncertainty and Climate Change

The Forum on Global Change Modeling met in October 1994 "to create a consensus document summarizing the debate on issues related to the use of climate models to influence policy" (Barron, 1995). Scientists now agree that average global surface temperatures have risen 0.5°C (1°F) since the 19th century. Highly probable changes due to an increase in anthropogenic greenhouse gases will include increases in global mean surface temperatures, increases in global mean rates of precipitation, rising sea levels, and changes in the biosphere. Nevertheless, substantial uncertainties remain in the magnitude and geographical distribution of these changes and in the rates at which they may be expected to occur. Climate models cannot at this point adequately determine changes in tropical storm intensity, predictions of local effects of climate change, changes in clouds, shifts in the intensity and distributions of precipitation, and changes in oceanic circulation (Barron, 1995; U.S. Global Change Research Information Office, 2001).

Resolving or reducing these uncertainties will require significant effort because of the inherent complexity and natural variability of the climate system. Improving model parameters to be able to predict regional effects more accurately will require a long-term commitment to research, observation, and careful reconstruction of past changes.

This uncertainty about the effects of climate change leads to conflicting attitudes toward what to do. On the one hand, a precautionary principle would focus on the concern that the effects of climate change could be large and dangerous; this perspective suggests that preventing climate change should take priority over other concerns. On the other hand, avoiding climate change will involve substantial costs; if the effects of climate change are uncertain, the very certain and very large costs might suggest delaying action until the uncertainty about effects is reduced.

In this case, explaining the effects of alternatives under the range of possible outcomes will influence decisions. Delaying action might have substantial benefits if the effects of climate change are not seriously harmful, but it might prove disastrous if climate change proves to have much worse effects than the average prediction. In contrast, reducing climate change now could, under the low-impacts scenario, reduce human well-being now because of wasted spending, but it could help substantially in the high-impact scenario. Understanding as much as possible the implications of both up-side and down-side risks will improve the basis of decision-makers' choices.

2.8 Performance Indicators of Each Criterion

Of course, each of these general decision criteria is likely to have a number of measurable indicators. That is, to evaluate how different alternatives do in solving a particular problem (e.g., reducing carbon monoxide concentrations), we need to have a measurable indicator or gauge. For instance, the physical/biological/scientific/technical effects of a policy could include such varied factors as the effects on air and water quality parameters (e.g., concentrations in parts per million, ppm), the number of acres of freshwater wetlands affected, number of people contracting a particular illness (e.g., asthma attacks), changes in recreational visitation, or minutes of travel delay per vehicle mile. Economic efficiency is typically calculated as a change in net benefits to society, but the distributional or social/cultural impacts of a policy can be examined for any number of different affected parties. Distributional issues can be summarized by whether the policy costs would be regressive or progressive, or by evaluating whether the benefits per household rise or fall faster than household income. Determination of operational practicality or administrative feasibility requires estimates of the agency budget required, number of employees, and an assessment as to whether the budgetary resources allocated to an agency are fixed or changeable. Legality differs depending on whether the decision-maker is a legislative body (which can pass new laws), an administrative body (which cannot pass new laws but can make new regulations), or a lower-level official. Uncertainties can affect many different parts of the policy questions and can be analyzed both separately and jointly.

To greater or lesser extents, all these factors are likely to play a role in an environmental policy debate. Perhaps the most important thing to notice is that the scientific impacts are only one set of criteria: the societal aspects also need to be considered. As discussed in Chapter 1, environmental and resource problems are not only science problems, but also human problems, and they need to be viewed through both sets of lenses.

The above list is likely to encompass most of the factors that might influence the decision on an environmental problem. An environmental impact statement (EIS) is likely to include many of these effects, with particular (but not exclusive) emphasis on the physical-biological-scientific-technical aspects of the problem. Executive Order 12291, issued by President Reagan in 1981, requires agencies, to the extent permitted by law, to use benefit-cost analysis in their regulatory processes. President Clinton's Executive Order 12898 on environmental justice requires agencies to evaluate the distributional consequences of its actions on minority and low income populations. Groups interested in a problem are likely to focus on subsets of these impacts: environmental groups are likely to concentrate on scientific effects, while those with a financial interest in a policy will pay particular attention to the distributional effects, and those who might implement the policy will focus on the administrative feasibility.

As a policy analyst you may have to determine impacts in virtually all these areas (for instance, if you are overseeing compilation of an environmental impact statement), or you may work on only one topic within one category of this list (such as considering the effects of a policy only on the interested party for which you work). As discussed in Chapter 1, you should make sure your analysis addresses the needs of your client (while maintaining your credibility as an analyst, of course). And, as discussed in Chapter 2, you need to define the amount of analysis you will conduct and the level of detail you will provide with your time and budget constraints in mind. For a given amount of time and money, you can cover a lot of topics at a fairly shallow, unsophisticated level, or you can do a very detailed and advanced analysis of one topic. Note, however, that the physical, biological and technical effects may have to be estimated so as to make possible calculating economic benefits and costs, even if these physical, biological and technical effects are not the primary focus our your study. Finally, a look ahead at upcoming chapters will show you some of the tools you might use to determine the effects of each alternative on each of your criteria.

Once you have identified alternatives for analysis, determined which criteria you are to analyze for each alternative, and conducted the analysis of impacts, the next step is to present them to your client in a form that will assist her in her decision-making capacity (if she has one). Sometimes your client will only want the information you have developed; in other cases, she might want it synthesized in a way that identifies which policies might be preferable. The *decision methods* to be discussed below suggest the range of ways that a decision-maker might approach the choice. It is unlikely to be your job to tell your decision-maker how to choose, but you can identify the impacts of different decision methods or rules on the preferred alternative.

3. DECISION METHODS

Once you have analyzed each of your criteria for all the alternatives, one highly effective way to display the results is through use of a matrix. Make a table where each row represents one criterion that you have analyzed, and each column represents an alternative. In each cell of the table, then, you can show the results of your analysis.

For instance, consider Table 3-1 below, where different ozone standards are being considered. Alternative A, the Status Quo, is the current ozone level in the region being studied; in language related to the National Environmental Policy Act, it is the "no action" alternative. Alternatives B and C are reduced ozone concentrations. Alternatives B and C further include sub-options, either using regulatory standards to achieve the goals or allowing marketable pollution permits.

This table presents some of the physical/biological effects (ozone concentration, effects on asthma sufferers, agricultural production losses); economic efficiency effects (additional cost to utilities, additional health and other benefits); distributional effects (agricultural effects, utility costs, asthma sufferers); and operational practicality/administrative feasibility (additional enforcement personnel required). If all these actions are equally permissible under existing law, equally socially acceptable, and unlikely to be affected by uncertainty in the analysis, then effects in these other categories need not be included, since they will not influence the decision. Thus reliance on the "with versus without" principle can often simplify the choice problem by eliminating from consideration criteria that do not vary across alternatives.

This matrix alone synthesizes a great deal of relevant information into a simple, easily understood format. A decision-maker can see the differential effects of the policies and can understand better the tradeoffs involved in the decision process. Your role may end with the presentation of these results.

In other cases, though, your client will want further synthesis of the information, based on guidelines for how she will want to make her decision. The following list provides some of the kinds of ways that decision-makers will want to put all this information together. It is unlikely to be your job to tell the client how to make a decision, but you may want to conduct several of the following analyses to determine whether the optimal policy depends on the decision method chosen.

Table 3-1. Matrix Approach to Relating Decision Criteria and Policy Alternatives

Decision Criteria/ Indicator	Alt. A, Status Quo	Alt. B1, Standards	Alt. B2, Permits	Alt. C1, Standards	Alt. C2, Permits
Air Quality Ozone Concentration	0.12 parts per million (ppm)	0.10 ppm	0.10 ppm	0.08 ppm	0.08 ppm
Health Restricted Activity Days due to Asthma	50,000	30,000	30,000	23,000	23,000
Farm Product Gains	0	2000 tons onions, 4000 tons grapes	2000 tons onions, 4000 tons grapes	3000 tons onions, 7000 tons grapes	3000 tons onions, 7000 tons grapes
Economic Costs to Utilities	$0	$40 million	$25 million	$120 million	$100 million
Economic Benefits Health, Ag, & Other Benefits	$0	$75 million	$75 million	$100 million	$100 million
Administrative Added Enforcement Workers Required	0	5	5	8	12

3.1 Dominance

Consider Alternatives B1 and B2, which achieve the same ambient ozone concentration; B1 uses regulatory standards, such as requiring all firms to roll back their emissions by the same percent, while B2 uses marketable emissions permits, which allow some firms to reduce their required cleanup in exchange for others abating extra emissions. According to the table, these policies have identical impacts, except in the cost category: under B2, costs to utilities are $25 million, while they are $40 million under B1. Since it is reasonable that higher costs to utilities are not good, then policy B2 is unambiguously superior to policy B1 – that is, B2 *dominates* B1. On this basis, there is no reason even to consider policy B1, since B2 is at least as good on all but one criteria and superior on the remaining one.

Under the Dominance rule, if one alternative equals or improves on another on all criteria, then the dominated alternative can be eliminated from the pool of alternatives. Why would it ever be chosen, if there is no factor on which it is superior to the alternative that dominates it?

The dominance rule for policy selection has the important advantage that there should be consensus about it on the part of all interested parties: why would anyone want to keep alternative B1 in the pool, when B2 is equal or superior on all counts? When it applies, it is a very powerful rule. The caveat "when it applies," though, must be taken very seriously. Rarely is one alternative dominated by another; the more typical case, as you will find by comparing any other alternatives presented above, is that each alternative has its advantages and disadvantages. For instance, the Status Quo option has the highest pollution, leading to the greatest health and agricultural effects, but it also has the lowest costs to utilities and for implementation. Though Alternative C2 (Permits) has lower costs to utilities than Alternative C1 (Standards), Alternative C1 requires fewer personnel for enforcement. No other alternative is dominated by another.

Dominance is worth checking for because it may help eliminate some of the alternatives from further consideration, and because some natural resources are used so inefficiently (e.g., irrigation water in the West) that we occasionally can identify dominant or win-win alternatives. Because it is rare that one alternative equals or exceeds all other alternatives on all criteria, the dominance rule will not often be a decision method that can be used to find the best alternative. Also be aware that if one alternative clearly dominates all the others on every decision criteria, there may be reason to believe a decision criterion may have been overlooked, or the other alternatives have been purposely designed to be inferior to the dominated alternative to make the dominant alternative the obvious choice. For rarely do solutions to environmental problems result in unanimous agreement that one alternative is superior. Perhaps the full range of alternatives was not considered, for instance, or the dominated alternatives were structured as straw horses, to be knocked down in favor of the preferred alternative. Alternatively, perhaps the only criteria presented were those that favored one alternative. If, for instance, "Additional Enforcement Personnel Required" were not listed in the above table, then Alternative C2 would dominate Alternative C1. Perhaps inclusion of a category for "Likelihood of Violation of Air Quality Standards" (suggesting effects of uncertainty) might favor Alternative B1 over Alternative B2, and make the dominance of Alternative B2 fade away.

The Dominance rule for decisions, when it can be applied, is very useful for eliminating substandard alternatives and reducing the number of remaining alternatives to be considered. As discussed, though, its usefulness is limited. It may in fact serve primarily to inform you that you have left something out of your analysis, either from the list of alternatives or from the list of criteria used to evaluate the alternatives.

3.2 Criteria Ranking

Sometimes a decision-maker has a well defined objective that permits relatively little discretion. For instance, the Clean Air Act requires that the primary National Ambient Air Quality Standards be set to protect human health, with a margin of safety; the costs of achieving that standard are not to be used in setting the standard. In that case, protecting human health is the primary criterion for choosing among alternatives; other criteria can only be taken into account in a secondary way. Since Alternatives C1 and C2 both increase protection of human health (as measured by Restricted Activity Days due to Asthma), by criteria ranking these alternatives are superior to the status quo or to either B alternatives, even if they are inferior on measures such as Cost to Utilities and Additional Enforcement Personnel Required.

Criteria ranking, then, is a rule that gives primary emphasis to one (or perhaps several) criteria, and the decision is made based primarily on that (or those) criteria. Other criteria need not be ignored; instead, they receive less attention. For instance, either Cost to Utilities or Additional Enforcement Personnel Required could serve as a secondary criterion, after human health effects, to decide between alternatives C1 and C2. The criteria ranking approach focuses the discussion on one or more criteria, and other criteria only come into play afterwards, perhaps as a tie breaker between comparable alternatives.

The criteria ranking approach has the advantage that the decision-maker must express very clearly her priorities. The criteria that are important to her, either due to legal requirements or to a guiding philosophy, must be stated clearly. A disadvantage of this approach is that it may provide little flexibility even in the face of very high opportunity costs. Even though the costs to utilities triple to quadruple in moving from the B standards to the C standards, a strict criteria ranking based on human health would not permit any trading-off of these increased costs with health. Sometimes a government agency is required to accept the lowest bidder on a contract, even if there are concerns about the relative quality of work or timeliness of the lowest bidder. Increased flexibility added to the criteria ranking approach would obviously give a decision-maker more discretion to take additional factors into account; if enough discretion is granted to consider those additional factors, though, then the decision rule may cease to be, strictly speaking, criteria ranking.

3.3 Benefit-Cost Analysis

Under a benefit-cost analysis criterion, economic efficiency becomes the guiding criteria or rule. The best policy should maximize net benefits to society – that is, provide the greatest net gain, without regard to who gains and who loses. After all, if the gains exceed the losses, then it is possible for the gainers to compensate the losers and, therefore, for everyone to be better off.

Under this criterion, it is clear that the marketable permit approaches dominate the standards approaches, since the benefits are the same for each ozone concentration but the costs are lower for the permits alternatives. Further, the 0.10 ozone concentration of Alternative B2 (permits) has net benefits over the status quo of $50 million, while the 0.08 concentration with permits has no net benefits over the status quo. Therefore, the 0.10 ozone concentration with marketable permits is the preferred alternative. Another way to see that the 0.10 concentration is superior to the 0.08 concentration is that the additional (marginal) benefits of going to the lower concentration are $25 million, while the additional (marginal) costs are $75 million.

Note that here, economic efficiency is being used as a decision *rule* rather than a decision *criterion.* As a criterion, the net benefits of an alternative would be a factor in the decision, but they would not necessarily determine the preferred outcome. As a rule, on the other hand, the net benefits decide the outcome. In essence, a benefit-cost decision rule is a form of criteria ranking, where the primary criterion is economic efficiency.

The first advantage of a benefit-cost decision rule is that it ensures that the benefits do outweigh the costs for an activity (in comparison to the status quo – remember the with-and-without principle!). While having positive net benefits may seem like an obviously desirable thing, many public projects are undertaken that impose net costs on society. For instance, many government activities provide subsidies to specified producers; typically, these are economically inefficient. A benefit-cost decision rule would eliminate those programs that cost more than they are worth.

Secondly, as noted above, because this rule ensures that society gains from the activity, then it is possible for everyone to gain from the activity. Even if some groups bear costs, the gains to the rest of society can be used to compensate them for the loss. For instance, in this example residents in general and agricultural producers benefit from the cleaner air, while utilities bear the cleanup costs. In this case, higher electricity rates from the consumers (and perhaps even higher rates for farms) might provide the additional funding needed for the utilities to pay for the cleanup. The higher rates, if less than the increased benefit from improved air quality, could make the utilities willing to pay for the abatement effort. In this case, everyone is at least as well off as before, and some gain, since there are net gains to be had.

These points also suggest several problems associated with a benefit-cost rule. First, and probably most important for environmental policy issues, *all* the benefits and costs of an action need to be incorporated into the analysis. Thus, in our example, it is important to include not only the market effects of ozone on agricultural crops, but also the (relatively) non-marketed effects on human health into the calculation. If ozone has other effects (as it does), such as reducing visibility and increasing corrosion of iron and other goods, then those need to be included in the calculations as well. Especially in the past, benefit-cost analysis often only covered those goods with easily measured market prices and quantities. Especially since the advent of the National Environmental Policy Act of 1970, economists have put a tremendous amount of effort into ensuring that environmental effects are included in benefit-cost analysis. These efforts are controversial – how much is an endangered species worth? Can a human life be measured in dollars? On the other hand, putting dollar values on these resources makes clear that these resources have real economic value, and that people are willing to pay to protect the environment. *Not* including them in a benefit-cost analysis biases the analysis toward environmentally destructive activities.

The second objection to benefit-cost analysis is that, even if it is possible to make everyone better off from an action, usually there are winners and losers from a government action. In the above case, agricultural producers and people sensitive to air pollution gain from reduced ozone, but utilities and their customers bear the costs. Is it surprising that utilities might oppose the tighter ozone standards? Is it surprising that agricultural producers and pollution-sensitive people will lobby for the 0.08 ozone concentration even though it is less efficient than the 0.10 standard? While benefit-cost analysis might be a useful assessment of the effects of a policy on society, the distributional effects of the policy are likely to have much more explanatory power for how interest groups behave. As long as decisions are made by people who might be influenced by interest groups, benefit-cost analysis may end up as a decision criterion rather than as a decision rule.

Box 3-2. BCA at EPA

Eban Goodstein (1995) evaluated two benefit-cost analyses (BCAs) from the U.S. Environmental Protection Agency (EPA) to determine whether EPA conducts good BCAs and whether BCAs affect regulatory decisions.

One BCA considered three options of varying stringency to determine acceptable levels of lead leached from solder on pipes into drinking water. EPA identified numerous health benefits from reduced lead exposure, but a high level of uncertainty limited EPA's ability to quantify health benefits. EPA failed to factor that uncertainty into the analysis and ignored benefits that could not be translated into dollars. Although EPA eliminated a clearly inefficient option, the uncertainty associated with benefits prevented EPA from distinguishing between two options with similar stringency. EPA ultimately selected the less expensive, and slightly less stringent, option.

The landfill BCA included 2 options for toxic leaching into groundwater: a less stringent "pollute and cleanup" option, and a stricter prevention approach. The benefits of access to suitable drinking water were reduced cancer risk and avoiding costs of replacing damaged groundwater. The cancer risk estimates were low because of uncertainties with the extent of leaching over time and human exposure to hazardous waste, and because few people rely on groundwater wells within a mile of landfills. EPA assumed that reduction of groundwater effects were all the benefits and insufficiently weighed uncertainty in benefit and cost predictions. EPA further considered a tradeoff between equity and efficiency and chose an option that seemed to favor equity for citizens residing near landfills over efficiency. EPA justified the selection of the more stringent pollution prevention option with unquantified benefits and equity implications.

Goodstein's analysis of these two BCAs led him to conclude that EPA overall conducts good benefit-cost analysis. EPA works to develop theoretically consistent benefit and cost measures, identifies and justifies assumptions, provides realistic upper and lower bounds for benefit and cost estimates, and includes background information. In neither case, though, did EPA choose the option that maximized net benefits, and Goodstein estimates the efficiency loss at $3 billion. On the other hand, the analyses may have enabled EPA to choose a more efficient option than would have been selected in the absence of the studies.

BCA does not offer significant benefits when EPA must address unquantifiable or intangible health benefits because it is difficult to identify efficient outcomes. Goodstein suggests that EPA consider devoting BCA resources to issues with large intangible and uncertain marginal benefits of regulation and to exploring alternative equity goals at lower costs.

3.4 Multiattribute Utility Approach

There is a reason that all the criteria presented in the table are included – they are all likely to be important to your client. The multiattribute utility approach is a way to include all these different factors into a decision rule. In its typical form, weights are assigned to each criterion, and the weighted criteria are summed; the alternative with the greatest weighted sum is the preferred alternative. ("Multiattribute" refers to the fact that multiple attributes (criteria) are included in the equation; "utility" refers to the economic model of consumer behavior, which assumes that individuals seek to maximize their utility, or well-being, by their decisions to consume a range of goods and services.)

For instance, suppose that the multiattribute utility function for this problem is:

Utility = [1000*(Onions, tons) + 8000*(Grapes, tons) + 1*(Health, Agricultural, and Other Benefits) - 1000*(Restricted Activity Days) – 1*(Cost to Utilities) – 50,000*(Additional Personnel)]/100,000

This equation says that the costs (shown by the negative weight) imposed by a Restricted Activity Day are worth 1000 relative to a dollar spent on utility costs. Meanwhile, a ton of onions (a benefit, because the weight is positive) is worth 1,000 dollars spent on utility costs, a ton of grapes is worth 8000 dollars of utility costs, a dollar of total benefits is worth a dollar of utility costs, and each additional enforcement hire costs 50,000. With these weights, the resulting values are:

Table 3-2. Illustration of the Multiattribute Utility Approach

	Alt. A, Status Quo	Alt. B1, Standards	Alt. B2, Permits	Alt. C1, Standards	Alt. C2, Permits
Value	-$500	$388	$538	$156	$354

With this utility rule, Alternative B2 is the best rule, followed by Alternative B1, Alternative C2, Alternative C1, and Alternative A.

Note that the numbers themselves are not necessarily meaningful; while the relative weights are important, the absolute values are not. The coefficients could all be divided by 1000, or multiplied by your age, or $500 could be added to this equation (which might be convenient, to make the Status Quo have a value of 0), and the same decision would result. In one case the specific numbers might be meaningful: if the coefficients were prices, then the results could provide the net benefits of each alternative, the benefit-cost rule. (In this case, care would need to be taken not to double-count some benefits and costs, such as including health effects both as Restricted Activity Days and as part of the Benefits figure.)

Obviously, with another set of weights, another alternative may score more highly. Thus, the values of the weights (relative to each other, as mentioned above) have a strong influence on the outcome of the analysis. Where do these weights come from? Here lie both the strength and the weakness of this decision rule. Once a set of weights is identified, then the optimal policy is easy to identify. The appropriate set of weights, though, is not typically published, and your client is unlikely to have a set of weights in her mind. You may have to work with your client to elicit the weights. You can first ask your client which criteria have positive values in her mind, and which have negative values. You can then ask about relative magnitudes: Is a restricted activity day more or less important than a ton of onions? Then you can ask for more specificity: e.g., is a restricted activity day worth twice as much as a ton of onions? Through this process, you can calibrate your client's values and use them in your analysis. Keep in mind, the resulting best alternative may be quite sensitive to your client's relative weighting. If other parties involved in this decisions have different weights, they may disagree on which alternative is best. Using the multi-attribute utility framework as a conflict resolution device might be one way to proceed in this situation: if various stakeholders attempt to develop a set of consensus weights, differences in opinion about what constitutes the best alternative will be brought out in the open through differences in opinion about the relative importance of each criterion.

Sensitivity analysis – changing uncertain parameters and seeing if the results change -- can also be very helpful in this setting. Does the determination of the best alternative change drastically if the weights change? For instance, suppose the weight on Utility Costs is changed from –1 to –2 (in other words, utility costs are more important). As Table 3-3 shows, Alternative B2 is still the highest ranked, but Alternative B1 is now second, followed by A, C2, and C1. In contrast, if a Restricted Activity Day is worth $10,000 of utility costs (ten times the previous value), then Alternative C2 is preferred, followed by C1, B2, B1, and A. Notice that, in all these variations, C2 is preferred to C1, and B2 to B1 (the latter expected by the dominance criterion). Also, it seems to require major changes in weights to make C2 more desirable than B2. Thus, if the weights in the original formula appear to be roughly right, the ordering seems relatively insensitive to change. If the orderings changed with only minor changes in the weights, then being more certain about the weights would matter much more.

Table 3-4. Illustration of Sensitivity Analysis in Multiattribute Utility

	Alt. A, Status Quo	Alt. B1, Standards	Alt. B2, Permits	Alt. C1, Standards	Alt. C2, Permits
2*Utility Cost	-$500	-$13	$288	-$1,044	-$646
10,000* Restricted Activity Day	-$5,000	-$2,313	-$2,163	-$1,914	-$1,716

3.5 Screening

In many policy situations, there are legal requirements that an alternative must achieve, or there may be consensus among key agents for a policy over some threshold values that must be achieved. The screening approach to decision-making takes those requirements and determines which alternatives meet all the screens required of a preferred alternative.

For instance, suppose that funding restrictions for the Environmental Protection Agency limit new enforcement personnel to 8, and that more than 25,000 restricted activity days of health effects is considered unacceptable. Alternatives A, B1, B2, and C1 meet the first screen (Alternative C2 falls out at that point), but only alternatives C1 and C2 meet the second screen. With both screens required, Alternative C1 is the only acceptable alternative.

The screening approach has the advantage of ensuring that a selected alternative meets key requirements. If these requirements are critical, though, it is reasonable to wonder why any alternatives that did not meet those requirements were proposed. However, many times it is impossible to know, without a great deal of quantitative analysis, whether an alternative will meet the threshold level of a particular criteria. If few or no alternatives meet all the screens, it is likely to be time to develop some new alternatives.

It is quite possible that more than one alternative will satisfy the screens. If this happens, then the screening approach does not provide a rule for how to decide among the remaining alternatives. Other decision approaches, such as criteria ranking, may be useful at this point to select among the remaining alternatives. Nevertheless, the screening decision method forces acknowledgment of legal, administrative, and key criteria when selecting among policies.

3.6 Mathematical optimization

As just mentioned, the screening approach is very useful for ensuring minimum standards are met, but it may have to be combined with other methods if more than one alternative satisfies all the screens. Mathematical optimization is an explicit way of combining these approaches. This technique combines a constraint set – a set of screens or other limitations – with an objective function – a goal statement. The goal statement, commonly referred to as the objective function, can take the form of criteria ranking – e.g., maximizing one value in the table – or it can be a multiattribute utility function, which incorporates a number of criteria with weights assigned to each one. The typical form of a mathematical optimization problem is to maximize (or minimize) the objective function, subject to achieving the constraints (the screens).

For instance, one objective function might be to minimize health effects of ozone, with constraints that total expenditures on ozone control not exceed $150 million, and that no more than 8 additional personnel are required. All the alternatives satisfy the first constraint – that is, the constraint "is not binding" on the choice of alternative. Alternative C2 does not satisfy the second constraint, since it requires 12 additional personnel; all other alternatives satisfy this constraint. Now that the screens have been applied and satisfied, by alternatives A, B1, B2, and C1, the objective function comes into play. Since C1 has the fewest Restricted Activity Days while satisfying the constraints, it is the preferred alternative using an optimization decision method.

Notice that, if the second constraint were changed so that more additional personnel could be used, Alternative C2 might become a feasible alternative; and, because its effects on human health are the same as C1, it might be a selected alternative. Since the level of the second constraint might affect the choice of preferred alternative, this constraint is said to be binding.

More complicated expressions can be used for either the objective function or the constraints. For instance, a multiattribute utility function can be used for the objective function, and constraints can be combinations of criteria. For instance, a constraint could require that total agricultural effects, defined as (the price of onions)*(tons of onions) + (the price of grapes)*(tons of grapes), plus total health effects, defined as (the cost of a restricted activity day)*(number of restricted activity days), not exceed some value.

Linear and nonlinear programming are often used for mathematical optimization programs. More about this technique can be found in Chapter 11.

Mathematical optimization can be complex to set up; getting an objective function and all the constraints can require tremendous research efforts (though advances in computing make the actual solution process much easier). Once the basic model is developed, though, it can be a very useful tool for simulations, for sensitivity analysis (such as examining the effects of changing either the objective function or the constraints), and even for developing new alternatives. The U.S. Forest Service has used linear programming heavily in its planning process for the National Forest System (Loomis, 1993), and it has been used as well for transportation planning, air pollution control modeling, and other applications. It is always important to keep in mind, though, that these models are only as good as the information that goes into them. If constraints do not accurately incorporate the workings of the natural or human environment, or if the objective function does not reflect the thinking of policy-makers, then the results cannot be expected to be very useful.

4. USING DECISION CRITERIA AND DECISION METHODS

We cannot emphasize enough the importance of distinguishing between decision criteria and decision methods. Criteria are the basis of the decision; a great deal of your work as an analyst is likely to be in identifying the effects of each alternative on the selected criteria. In and of themselves, though, they do not determine which alternative is superior. Even when determining the effects on the criteria is a simple exercise, the decision itself may be complex and difficult, if no alternative dominates all others.

As should be clear by now, rarely is it obvious which alternative is the best one. Different decision methods, or even the same decision method with different weights or values, will change which alternatives are preferable to others. While it would be wonderful to come up with an algorithm that determines the optimal policy, there is a reason that decision-makers are paid well (in power and prestige, if not financially). These decisions are rarely easy; an easy decision would require either that one alternative dominates all others, or that there is a clear societal consensus on how to weight criteria in the decision. The former, as mentioned over and over, is likely to occur if alternatives are narrowly formulated; the latter would take a miracle.

Your job as a policy analyst, as we discussed back in Chapter 1, is to provide your client, a decision-maker, with information to assist her in making her decision. Thus, your job is likely to involve assembling decision criteria and presenting them to your client in a form that she will find useful. Your job may include analyzing the alternatives according to one or more of the decision methods just discussed. As we discussed back in Chapter 1, though, we do not expect that your job as policy analyst will be to make the final decision. Still, tasks such as developing an appropriate list of criteria, formulating alternatives, determining the effects of each alternative on each criterion, and determining the effects of different decision rules on the choice of alternative will help your client to make the best decision possible, not to mention keep you busy.

Chapter 4

Steps in Conducting an Environmental Policy Analysis

What do you do when you conduct a policy analysis? In many ways it bears a striking similarity to the research papers you probably did as a student. As with any research paper, your objective is to assemble information in a readable form and to document your methods and conclusions. Because your policy analysis is likely to be more widely read and taken more seriously than most term papers, you probably want to be more systematic about how you conduct your analysis.

This chapter will provide you with a set of steps that you should take in conducting a policy analysis. It is quite likely that you will face these steps in any analysis that you do, though perhaps not in the order specified. Of course, rather than view these steps as part of a checklist, you should adapt them to your own style, to the policy question at hand, and to your client's needs.

1. STEP ONE: DEVELOP A RELATIONSHIP WITH A CLIENT

Chapter 1 has already discussed the role of the client in policy analysis. Whether your client is your employer, an organization for which you are doing contract work, or some other relation, you should have someone in mind when you are conducting your study. Having a client is critical for defining your analysis topic (Step 2), and everything else is derived from that step.

2. STEP TWO: DEFINE YOUR ANALYSIS TOPIC OR PROBLEM TO BE ADDRESSED

In many ways, this step cannot be separated from Step 1. As discussed in Chapter 1, what question you seek to answer depends strongly on the client for your results.

This step involves several components. First is determining what is the key policy issue the client must answer or problem to be solved. Clearly, your analysis must address this key policy issue head on. You must familiarize yourself with the context and the key components of this issue. For example, if the issue is how best to meet a given ambient carbon monoxide standard, then you must understand the sources in your area, the current level of this pollutant, and the level of the standard. Next, you must determine what information about this issue is critical to your client's decision. This question is the core of this step. If your client is the U.S. Environmental Protection Agency (EPA), the focus of your analysis of meeting the carbon monoxide standard is likely to be different than if you are working for the American Automobile Association.

As discussed in Chapter 1, you should also consider the possible consequences of your results: what if your results do not support the perspective of your client? If your client is likely to be searching only for results that support her perspective, you will want to be honest with your client about the possibilities that certain types of analysis or specific issues may not support her preconceived position.

Next, you must identify the baseline for analysis. That is, what does your client consider to be the status quo without any action, the "without" scenario? Because, as discussed in Chapter 2, you will be comparing the effects of any policy to the "without policy" alternative, you need to decide what this baseline will be.

Finally, you should determine the overall study constraints in terms of time, personnel, and budget available. As also discussed in Chapter 2, your analysis will be different if you have a great deal of time and budget than if you have to scrape together your results in one day.

3. STEP 3: DESIGN THE ANALYSIS

Now that you have your question identified and your time and budget constraints set, how will you actually do the work? While the following discussion must be tailored to the specific project you are doing, the following components are likely to help you in this step.

First, review the existing literature related to your policy question. How have people approached your question in the past? What useful information, particularly about methods and modeling approaches, and perhaps about data availability, can you learn from existing studies?

Next, determine the likely future without the project or the consequences of the no action alternative. Again, because your analysis of the problem must involve a comparison to the baseline, now you must assess what the world would look like if the baseline continues.

Next, you should determine what methods you will use for your analysis. Does your client want a benefit-cost analysis or does she want to focus on employment effects using an input-output model? Will you use simulation methods to identify the impacts of various policies? Will statistical techniques help you to identify a critical relationship, or will you instead need to use a linear programming approach to find a cost-minimizing solution? Chapters 6-11 will help you understand how and when to use some of these techniques.

Finally, you need to identify your data requirements. Your data needs will be based on your review of existing literature, the choice of analysis approach (e.g., benefit-cost analysis versus an input-output model), your monetary and time constraints, and the data already available. At this step you will need to decide whether there are sufficient existing or secondary data, or whether primary or original data must be collected. In either case, you must identify the geographic and time scales for your data. That is, what are the relevant units from which to collect data (firms, households, city averages), and over what geographic range (all cities in the state, the country, or the world)? If time series data are to be used, you must determine how far back in time you will go. Collecting data further back in time increases the sample size, but the possibility of less temporal relevance increases as one goes back in time to periods where other factors may be quite different than today. For instance, for a study of telecommuting, collecting data from before the 1980s might be useless, since very few people could work from home before that time period.

4. STEP 4: DATA COLLECTION

In the previous step, you decided what data you will need for your analysis. Now you need to collect it. This may involve trips to a library, going through your client's records (paper or electronic), downloading data from a web site, purchase of data from other sources, as well as design and implementation of surveys or interviews. Chapter 2 discusses some of the advantages and disadvantages both of using existing data and of collecting new data. This step will also rely on the agreement you and your client made on your time and budget for your work.

5. STEP 5: DEVELOP AND SCREEN ALTERNATIVES

This step may be carried out earlier, when you are designing the analysis; or, you may come to it later, after you have done some preliminary data analysis; or, you may want to reconsider your list of alternatives at all these stages. In this step you will determine what are the possible solutions to the problem identified in Step 2. What options or choices does your client or the decision-maker have? (e.g., approve building a new highway to reduce traffic congestion). What options might be available to other decisions makers? (e.g., banning heavy trucks from highways at rush hour, road pricing to reduce traffic congestion). Are there additional options that might make more sense? (e.g., asking employers to allow telecommuting to reduce traffic congestion). This step can be very important: you may be defining the breadth of choices for your client. As discussed in Chapter 3, you want to have a wide range of alternatives that covers the spectrum of options. The National Environmental Policy Act requires a wide range of alternatives be analyzed, even those the "action or lead agency" may not have the legal authority to implement itself. Since the alternatives will define the universe of possible solutions to the issue, you want to ensure that stakeholders have their options taken seriously. At the same time, if you present too many options, you may overload your client with choices, and her ability to review them will be reduced.

How widely do you wish to think of alternatives? For instance, should you consider the current budget for a program fixed, or is there the possibility of increasing or decreasing it? Can you consider changing a current pollution standard, or should your analysis be limited to different ways of achieving it? Some alternatives may be limited by your problem definition in Step 2; in some cases, developing your alternatives may involve some reworking of that problem statement.

These alternatives will be the core of what you will analyze. Based on the alternatives that you develop, you may need to revisit some of the previous steps (such as data collection) to ensure that you can complete your analysis.

6. STEP 6: PERFORM THE ANALYSIS

Now you will actually do the work outlined in Step 3. You will want to estimate and then compare the effects of all your alternatives for a common set of impacts (see Chapter 3 for a suggested list of impacts that you may wish to measure). Don't forget to include the baseline or "no action" alternative, either as an explicit alternative, or as the basis for comparison of other alternatives. Various models, statistical techniques, or even expert opinion may be used to predict or estimate the consequences of implementing each alternative on each criterion.

One way of organizing the resulting predicted effects of the alternatives, as discussed in Chapter 3, is to develop a matrix: all the alternatives will be on one axis, and the measured effects will be on the other axis. Filling in such a matrix ensures that all alternatives are evaluated according to the same criteria.

You may also want to compare the advantages and disadvantages of different alternatives. As discussed in Chapter 3, there are a variety of different ways (decision methods) to assess alternatives; which method, if any, that you use is one of the questions you will have wanted to determine when you defined the problem in Step 2.

You may wish as well in this step to perform some sensitivity analysis. Sensitivity analysis examines whether your results are likely to be very similar to their current form if some of your underlying assumptions or data change. If the discount rate changes by 2%, for instance, will the same alternative still be superior? If you have underestimated (or overestimated) the effects of a policy on pollution, would you still want to recommend the same alternative? If the result does not change as your underlying assumptions change, your result is said to be *robust*. If the result is highly sensitive to your assumptions (hence the term "sensitivity analysis"), then you may wish to put more resources into determining the precise levels of your variables, or perhaps your preferred alternative will be one that is more or less variable across different assumptions.

7. STEP 7: PRESENT RESULTS

Of course you intended to present the results to your client. In the process, do not forget to make the results understandable to her. Here, knowing your client (and understanding how much she knows about the topic) will help you to present the information in the most useful way possible. Do you need to provide her with background on the issue, or is she already well versed in the topic? Does she want a one-page briefing memo, with bullets highlighting the key findings, or does she want to see background on how you did the analysis, and why? Will charts or tables help to summarize the information? Do you need to explain some of your terminology, or can you eliminate jargon altogether and replace it with common language?

Additionally, you want to find a way to ensure that she understands the merits (and problems) of each alternative. While you may have compared the alternatives, and perhaps even made a recommendation, she needs to be able to determine for herself which alternative best meets her objectives or solves the problem you attempted to address.

How you present your results can have a significant impact on what people will learn from your analysis. Results in graphs and tables are often more easily accessible than results discussed in text. In part because they are more easily accessible, results in graphs and text can also be easily taken out of context or misinterpreted. If the uncertainties surrounding your results are large and important, you may want to graph the uncertainties or provide a range in a table rather than just the average prediction; if a table does not include some of the nuances associated with your results, you may want to take those results out of the table and discuss them in the text, surrounded by those nuances. It is becoming more common place to have your results reviewed by others before finalizing them. This can be as informal as an in-house review or as formal as an outside peer review. The extent of review will depend on the desires of your client and how certain you are of your analysis. Thinking in advance about the implications of how you present your results can increase the likelihood that your results will be understood as they should be.

8. HOW TO USE THESE STEPS

As you plan your policy analysis, we encourage you not to think of these steps as being stairs that lead you to the top, where your completed report awaits you. Instead, these are different stations that you should visit along your way. You may need to go back and forth between some of them several times. For instance, when you perform your analysis (Step 6), you may want to re-evaluate your set of alternatives (Step 5), trying out some new ones that arise from your assessment. Conducting sensitivity analysis (Step 6) may lead you to refine your data (Step 4). Discovering what data are available (Step 4) (with luck, that more and better data are available than you thought) may lead you to revise your plan for your analysis (Step 3). The point is allow time in your study design to iterate through these steps a couple of times, rather than viewing them as a once and for all, linear path.

As you become more experienced in policy analysis, you will refine how you implement these steps for conducting the analysis. In addition, you will become quite adept at tailoring the steps to meet the wide range of policy issues you will be called upon to analyze. We hope that this list will help you come up with your own style for implementing these basic steps.

Chapter 5

Modeling

Much of what you will do as a policy analyst is to predict the consequences of alternative policy proposals before they are implemented. Prediction, by definition, requires an extrapolation from the current situation to a new one. If it were possible to create a reality identical to the current one, impose the policy, and compare that alternative reality to the results of the existing one, you would know exactly the effects of the policy. Until we have replicated our existing societies in new locations (new planets?), we will have to rely on other methods to predict the impacts of a policy. Unavoidably, predicting these impacts will usually involve models.

One way of making a prediction is relying on your personal expertise and extrapolating from historical conditions. However, there are two limitations with this approach. First, a new analyst may not have years of experience with the geographic area, subject matter, affected populations, or other matters for which responses to a policy need to be predicted. Secondly, the general public, the courts and public officials typically want a well documented rationale for the prediction. When critics of your analysis propose an alternative prediction, you want to have a strong basis on which to assess their findings and compare them to yours. Models are commonly accepted procedures to make predictions about the effects of activities without actually undertaking the activity. At the same time, you must always keep in mind that a model is no better than an approximation of reality. Just as Newtonian physics was shown to be a good but not perfect approximation with the advent of relativity, any model of how the world works should be considered an approximation that will not always be exactly correct.

1. DEFINING A MODEL

A model is an abstraction and simplification of reality that can help explain why phenomena in the world happen, predict what impacts will result from various behaviors, or both. In some cases, your work will focus on the explanation function of a model; in others, the prediction matters more than the explanation. Sometimes the same model can be used for both purposes; in other cases, the two functions are not compatible. Most importantly, while you cannot avoid using models, you must also learn the limitations of whatever model you are using. Any model, because it is a simplification, cannot incorporate all of reality. In choosing a model, you need to determine whether any important aspects of reality are left out of the model, or whether the simplifications in the model will not affect your use of it.

In your everyday activities, you use models of physical and human behavior to get along. When you walk to an intersection with a stop sign and want to cross the street, you predict that drivers will stop at the sign and give you a chance to cross, based on a model that says that most people obey stop signs when driving. (You are also likely to look carefully when you cross the street, because you know that your model does not work in all situations.) You use maps to figure out how to get from one place to another: the map serves as a model of the landscape you seek to navigate. At the same time, a national map of the U.S. will not serve you well as you try to find a particular street address in your area. Just as different kinds of maps are useful in different situations, models typically work better in some situations than in others. As you work with models, you must learn when a model will provide you the information you seek, and when it is inappropriate.

The example of maps can illustrate the explanation and prediction function of models. You can use a local road map to predict how to get from one place to another. A larger-scale map, with more topographic features, may be useful to help explain why roads are located where they are (for instance, to avoid going through a mountain range, or to connect major cities). A topographic map of a mountainous area, with its indication of slope steepness and aspect (direction), may help predict where avalanches are likely; at the same time, underlying this use is a model that states that avalanches can be explained in part by slope and aspect.

Models come in many shapes, sizes, and purposes. They can be very simple, such as a rule of thumb saying that most drivers stop at stop signs, or they can be very complex (such as models used to predict weather or ocean currents). Many phenomena can be modeled using either a rule of thumb or a very complex relationship. For instance, one rule of thumb might be that a 100 percent increase in the price of gasoline will lead to a 10 percent reduction in gasoline use. A more complex model might examine how this relationship is affected by income, location, the availability of mass transit, or political affiliation. In some cases the rule of thumb might be adequate for your purposes; in other cases, the more complex model might be needed to account for specific circumstances.

We often evaluate ("validate") the quality of models through their ability to predict and explain the phenomenon of interest. Typically, the model will be run (often many times) with assumptions that reflect a reality with a known outcome, and the results of the model are compared to what really happened. For instance, if gasoline prices increase 100 percent, we can later evaluate whether gasoline purchases decreased by 10 percent. If this simple model also explains how gasoline consumption will respond to different percent changes in price (e.g., a 20 percent decrease in price will lead to a 2 percent increase in consumption), then this simple model may well be sufficient for modeling, for instance, the effects of a gasoline tax.

Often increased complexity of a model gives it the air of superiority. After all, inclusion of more data and more interrelationships must provide superior insights than a simple model with less information. This attitude toward more complex models is not always justified. Sometimes complex models are so specific to a particular situation, such as a particular locale or time, that its use outside those circumstances will provide completely misleading answers. It is also possible that erroneous relationships can be buried somewhere in the model; if the model is not validated sufficiently, those erroneous relationships may not be found until the model produces unbelievable or wrong predictions. Simple rules of thumb can perform better than very complex models applied to situations beyond those for which they were designed.

2. TYPES OF MODELS

Models can take many forms and serve many purposes. The range includes everything from a descriptive story or a single equation to a matrix or more complex optimization models. They can be equations, graphs, computer programs, or physical replicas (such as scale models). Models are created, studied, and used in virtually all academic disciplines, since all disciplines are involved, in one way or another, in trying to find simplified explanations of aspects of our world.

Models can be theoretical (meaning that relationships among variables are discussed without necessarily being linked to data) or empirical (meaning that data are involved). Theoretical models are typically used to suggest hypotheses about how systems behave; empirical analysis can then be used to test whether the theory is correct. Both are necessary for each other:

theoretical modeling suggests what variables and relationships are important for the empirical work, and empirical modeling both tests the model and determines the magnitude and importance of various effects. Usually decision-makers want quantitative empirical results whenever possible. The empirical models that you use to provide those results, though, should be based on theoretical models that stand up to empirical scrutiny.

Models can be positive (descriptive) or normative (prescriptive). A positive model will describe the outcomes from a certain set of practices, but it will not determine whether that outcome is desirable or not. A normative model is based on a decision rule (see Chapter 3) and will determine, among various alternatives, which one is superior. Sometimes the difference between these two types of models is not very strong. For instance, a linear programming model (see Chapter 11) is based on maximizing an objective function while satisfying constraints. This form can be positive – for instance, it could be used to determine the maximum amount of timber that could be harvested while ensuring protection for different amounts of wildlife habitat, thus tracing out a relationship between timber harvesting and wildlife habitat protection. Alternatively, the model could be normative, where the objective is to maximize timber harvests while protecting a specified amount of wildlife habitat. It is important to keep this distinction clear in your head while modeling. If you are building a normative model, on what basis will your model decide the superior outcome? Because modeling forces an explicit statement of your assumptions (see below), a normative model will require you to state exactly what you consider the objective of your analysis to be.

All the topics described in the second part of this book are empirical methods involving models. Often these models have embedded within them other models. Environmental policy analysis typically draws from scientific models of topics ranging from air pollution transport to wildlife-forage relationships. For instance, a benefit-cost analysis of protecting an endangered species through habitat protection must include some method of estimating the relationship between habitat protection and the endangered species. For this reason, environmental policy analyses usually require some ability to understand and work with either researchers or the results of research from a range of different disciplines.

3. ROLE OF ASSUMPTIONS IN MODELS

Model-building will require you to make a number of assumptions. Indeed, forcing you to think about the assumptions underlying your analysis is one of the more useful functions of models. If the assumptions can be justified from previous research or experience, your model will be more credible. If the assumptions rest on shakier ground, you may either want to recommend that more work be done to examine those assumptions, or you may want to perform sensitivity analysis to determine whether your results change with small modifications in those initial assumptions.

Assumptions take several forms in model-building. First, assumptions identify the key variables that influence the system under study. As noted above, these variables are frequently identified from theoretical work and verified through empirical studies. In a global sense, everything in the world affects everything else. For modeling purposes, though, you can probably get away with focusing on a few key variables and eliminating much of the rest of the world. For instance, suppose you are estimating water pollution from agricultural practices. Certainly hundreds of things might influence water pollution levels, but for a workable model capable of evaluating alternative agricultural practices, a handful of those hundred variables will probably give you a very good prediction. While your prediction will never be perfect, even with 100 variables, it will likely be much better than a "best guess" of the relationships. By explicitly identifying your data needs, you will put bounds on how much data you need to collect, while ensuring that you do know what data you need to get.

Second, assumptions help you identify how those variables are related to each other. Of most importance is determining what variables are determined by other variables – that is, what variables are endogenous (determined by others), and which variables are exogenous (determined by factors outside the system). Exogenous variables are used to predict endogenous ones. For instance, steel production (the endogenous or dependent variable) is determined from amounts of raw materials (iron), labor, capital equipment, and energy (the exogenous or independent variables). Once these variables are set, your model will predict the amount of steel that can be produced.

Variables are not always endogenous or exogenous. For instance, in a model of individual shopping behavior, it is probably appropriate to consider prices exogenous, since no individual can affect them; on the other hand, in a market system, the overall market determines the price (and quantity) of the good, and both those variables are endogenous. As will be discussed in Chapter 6, understanding the difference between exogenous and endogenous variables is critical for conducting and interpreting statistical analysis. If you confuse the two kinds of variables, you may end up with misleading, inaccurate results from statistical work.

Third, assumptions provide information as to when the model is applicable. That is, if the assumptions of the model are met, the model should provide a reasonably accurate prediction and explanation of the phenomenon under study. If the assumptions are closely approximated, the model still may be useful. If in fact the assumptions of the model depart dramatically from the case at hand, the model should not be used for that case. In this case, if the model is used it would not be surprising if its predictions were far from accurate. Chapter 2 provides some of the methods, advantages, pitfalls, and examples of using data and models outside the situation in which they were developed.

Box 5-1. The Importance of Checking Model Assumptions

The U.S. Environmental Protection Agency (EPA) released an inventory report that estimated the United States' anthropogenic greenhouse gas emissions and sinks from 1990 to 1996. This inventory was intended to be a tool to evaluate the feasibility of possible greenhouse gas reduction strategies (U.S. Environmental Protection Agency, 1998a). The levels of one greenhouse gas, nitrous oxide emissions, raised some concern. The inventory provided estimations that nitrous oxide emissions, of which about 2% originated from the catalytic converters in cars and trucks, contributed almost 7.2 percent of the global warming gases (Wald 1998). This report implied that catalytic converters reduced ground-level pollution while exacerbating global warming by releasing large quantities of nitrous oxide.

Puzzled by this conclusion, EPA further investigated the data and traced the problem back to the original model. The Agency had used data from the Intergovernmental Panel on Climate Change (IPCC), a well-known and respected panel of climate change experts. Given the limited amount of information regarding catalytic converters and nitrous oxide emissions, the IPCC chose to use the best available data. The best available data were from a French study that had been conducted under circumstances that differed from those considered by the EPA inventory. Furthermore, the French study sample size was very small: most of the data were based on a study that used one car (U.S. Environmental Protection Agency, 1998c). Consequently, EPA gathered additional information suitable for the inventory and revised the nitrous oxide emission factor. When you are using others' data or results for your analysis, it can be very important to know the source and limitations of those data and results if you want your results to be accurate.

4. RATIONALE FOR USE OF MODELS

There are several advantages of models to both the analyst and the decision-maker. To you the analyst, models provide a means to make a very complex system (natural or man-made) more understandable. By increasing your understanding of the system, you can design better policies to bring about desired changes, or make more accurate predictions about the future.

This abstracting from reality helps prevent you from losing sight of your basic research question.

Models also provide a transparency that can assist in building agreement among different groups interested in a policy. Frequently (but not always!), those on different sides of an issue perceive a benefit from good scientific understanding of a phenomenon. For instance, better understanding of the factors that lead to urban air pollution would be supported by both government agencies (who want to know what to regulate to improve air quality) and environmental organizations (whose interests in this case might well align with those of the agencies), and probably by some polluters (well, at least those who feel that their contribution to urban air pollution is overestimated). Even those who could be targeted for increased regulation might push for better models, by questioning the models being used and working to promote greater accuracy. The overall result of the process could be better understanding of the science of urban air pollution.

Models are very useful for simulating the effects of alternatives without actually implementing them. For instance, a model of water pollution from agricultural practices can be used to predict the effects of a restriction on fertilizer use. A model of the effects of timber harvesting on deer in Wisconsin can be used to examine whether different timber management regimes affect the size of the deer herd. If the model is based on the best available scientific knowledge of the system being studied (or a not-unreasonable approximation of that knowledge), then the focus of discussion among interested parties will not be on the effects of different alternatives, but the relative merits of different alternatives.

One of the great advantages of using a model is that it can provide consistency in analyzing alternative policies. For instance, suppose you are using a model that predicts water pollution resulting from farmer applications of fertilizer, and you are examining the effects of different fertilizer management regimes. You will probably be more justified in comparing the effects of the different regimes if the results all come from the same model, rather than if you used different models for different regimes. All models have some error in them. You are likely to have better comparisons when the form of the error is constant across your results than if different models are used: in the latter case, you are not sure whether you are comparing differences from the alternatives, or differences from the models. On the other hand, if your model is not designed to handle all the regimes – for instance, your model works well for corn but not for tomatoes – then your predictions are subject to error regardless whether you use the same model for corn and tomatoes, or different models for corn and tomatoes. Again, you need to be careful about whether you are applying your model to a situation for which it is appropriate.

Additionally, models provide evidence that results, policies, and plans were arrived at in an objective, scientific way. Unfortunately this aura of scientific management is provided by the model even when the model is not used appropriately, is not well validated, or has other significant flaws. If the model is available for review from all affected constituents and outside experts, and if it is subjected to testing of its validity, the scientific basis for the model will be justifiably enhanced.

5. ISSUES TO CONSIDER IN BUILDING A POLICY ANALYSIS MODEL

Conducting a policy analysis may not require you to build detailed models of the phenomena you seek to study; indeed, in many cases, your job will be primarily to use existing models and their results. Nevertheless, as you synthesize information from various sources, the modeling issues described below will help you to put all the information together in a consistent, well justified analysis.

5.1 Problem Definition

Chapter 1 has already discussed the importance of problem definition, and Chapter 2 has helped you think of the time, budget, and other constraints that you face in conducting your analysis. The following issues must be addressed in building a model. You need to ensure:

(a) that your model answers the questions that you need answered. For instance, if you need the costs of an alternative, a purely ecological model will not serve your purposes.

(b) that you will have sufficient time, money, and expertise available to build a model sufficient for your needs. Modeling can be very expensive and time-consuming, or a combination of rules of thumb and existing research can be pulled together fairly rapidly and cheaply. Either approach may be appropriate for the purposes of you and your client, as long as you and your client have negotiated these aspects in advance.

(c) that you have thought about both the current use of the model and possible future uses. If you expect that your model will be used only once, your model can be made highly specific to the conditions of your current study. If you think your model will be used again in the future, you will want to build it so that it can be applied to multiple situations.

(d) that you have thought about the consequences if your model is wrong. If your model is assessing impacts of a major, expensive activity with possibly irreversible environmental and social effects, then you are likely to want to spend more time and effort on your model to make it as accurate as possible. If the consequences of the model predictions are reversible (for instance, farmland might be brought out of production temporarily), then the need for being precise is much smaller, and you (and your client) could save money on the modeling effort.

5.2 Model Scale

At what level should your model be developed? Will it be applicable to an individual, a neighborhood, a city, a county, a region, a state, or nation, or the world? Will it work for some pollutants but not others? Will it work for water pollution but not air pollution? Will it apply to all ruminants, or only to deer? A model built for one scale may not work very well when applied to a different scale.

If you are using an existing model rather than building one, this issue can be especially important. Does the model you propose to use apply to your situation? You may want to do some validation of the model to determine if it is appropriate for your project.

5.3 Formulation of the Model

The assumptions that you plan to use in modeling need to be made explicit. As discussed above, you will want to draw on theoretical literature to determine what variables are important, what relationships these variables have to each other, and when this model will be applicable.

5.4 Data Collection

Earlier in this process you identified what variables are important. Sometimes you will need to collect original data using surveys or interviews; in other cases, you can rely on existing data sets. In either case, you need to get data that match what you need as closely as possible. Like model building, data collection can be time-, expertise-, and budget-intensive, or it can be done with very little of any of those.

The quality of data depends not only on the scientific rigor with which it was assembled, but also its applicability to your question. If you need data on how elk respond to timber harvesting, will data on deer, no matter how carefully collected, really serve your purpose? If you need household income for a city, will per-capita income for the state be an appropriate proxy? Again, you need to consider scoping questions as you collect your data.

5.5 Model Estimation or Calibration

For a statistical model, this is the point where you will use the data to estimate the coefficients in the model and/or apply the model coefficients to make predictions. For a simulation or programming model, you will assemble the equations and their relationships to each other and run the computer simulations. Considerations such as whether linear or non-linear functional forms are superior will often be tested at this point. In any case, you want your model to be the "best fit" for your data that you can achieve.

5.6 Model Validation

How do you know if your model does what it is supposed to do? Models are usually validated by running them for situations with known outcomes, and comparing the model's results with what actually happened. If you cannot conduct validation of this sort (due, for instance, to lack of data for comparison), peer review of your model can give you useful feedback. For multiple regressions, a variety of "goodness of fit" tests can be performed as well (see Chapter 6).

The results of your validation efforts are almost certainly not going to perfectly replicate the real outcomes; how similar they should be depends on how precise you need your model to be (Step 1). In some cases you will not be satisfied unless they are almost the same; in other cases, if the results are of the same order of magnitude, or even the same direction, you will be satisfied. If your model results do not satisfy you, you will need to check your assumptions, your data, and your model estimation to try to figure out where errors lie.

If you have concerns about some of your assumptions or some of your data, part of your model validation could include sensitivity analysis of these factors. If the results do not change much, the model is said to be robust to the different assumptions. If the results do change more than you desire, then you should probably put more time and effort into ensuring that you have these factors as correct as possible.

Once your model is performing to your satisfaction, you can now use it for the functions for which you developed the model. Most likely, you will use your model to predict the consequences of various alternatives to fill in part of a decision matrix (see Chapter 3) or as a decision rule to select the preferred alternative.

6. PROBLEMS THAT CAN ARISE WITH MODELING

Problems can arise from any of the steps described above. You may oversimplify the problem due to limited time, budget, or expertise; you may not be able to apply the correct model specification due to lack of appropriate data or computational ability; you may use a model inappropriate for the scale of your problem (for instance, using data from one neighborhood to represent that of the entire state); you may not have the right data, or your data may not be very good; or your model construction may omit critical relationships or describe those relationships poorly.

At the beginning of this chapter, we described modeling as a simplification of reality. Any simplification is imperfect. Your goal as a developer or user of models is to get a sense of when your model gives you useful information of the degree of precision required by your client, and when your model will not serve this purpose. You will never succeed in being able to replicate reality exactly. If you can provide information that provides a sufficient sense of the outcomes of different policy alternatives, though, you will have done your job well.

7. CASE STUDY: MODELING ENVIRONMENTAL INEQUITIES

Are poor or minority populations disproportionately affected by environmental harm, such as exposure to toxic waste sites or hazardous air pollutants? If so, why? This policy question has received increasing attention from activist organizations and government officials. For instance, President Clinton issued Executive Order 12898 in February 1994, requiring federal agencies to consider the effects of their activities on the distribution of environmental impacts in the population.

From an activist's perspective, identifying a correlation between environmental hazards and poor or minority populations may be sufficient to promote action. A policy with discriminatory impact – that is, a policy that affects different groups of people differently, regardless of the intent of the policy – may be considered seriously flawed, wrong, or immoral. If current public policy or private behavior leads to an inequitable distribution of environmental hazards, then new policies need to be put in place to protect these groups.

On the other hand, from a policy maker's perspective, the problem might be easier to address if the underlying causes are identified. Identification of causes can lead to proposed solutions, targeted to those problems. Helfand and Peyton (1999) identify four hypotheses for why this discriminatory impact may exist:

7.1 Income effects

Perhaps the fundamental problem is poverty, especially since minority groups in the U.S. disproportionately have low incomes. Income effects can show up in several ways. First, poor people live in areas with low property values (since they cannot afford to live in areas with high property values); environmental hazards, such as hazardous waste sites, gravitate toward areas with low property values as well, to reduce disposal costs. Secondly, poor people may be more willing to live near facilities that produce environmental risks in order to get jobs associated with those facilities. Third, if people's demand for a high-quality environment increases with income, then poor people will have less resistance than wealthier people to having environmental hazards in their midst.

7.2 Education or information effects

Poor and minority communities frequently have lower educational opportunities and attainment. They may have less access to information about the hazards in their communities, or they may not have the background to understand the risks fully. In either case, they will show less desire to avoid these risks than others with better information or education.

7.3 Political access or power

Poor and minority groups frequently have less ability to influence political processes than wealthier, majority groups. As a result, if government decision-makers influence the siting of these facilities (through zoning or tax programs), the facilities may end up in poor or minority communities because those groups cannot or will not protest as effectively.

7.4 Discrimination

Here, government decision-makers or siters of polluting facilities specifically target poor or minority communities (or specifically avoid other communities) regardless of other factors.

A large number of empirical studies have tried to determine if there is a correlation between poor or minority populations and environmental harms. The results depend a great deal on how the problem was modeled, and what set of data was used. Finding a correlation may be satisfactory if establishing a discriminatory impact is sufficient to lead to changed policy. However, in other cases discriminatory intent – a deliberate plan to harm some groups while leaving others unaffected – may be required. If so, then correlation is insufficient for action to result, and a model must be able to distinguish whether environmental inequities result from discrimination or from one or more of the other causes mentioned here.

As is well known in statistics, "correlation is not causality." Given that there are multiple independent factors influencing the siting of a polluting economic activity, a multi-variable model that lends it self to multiple regression (Chapter 6) is needed.

For purposes of illustration, the variable we wish to predict and explain is the siting of stationary sources of air pollution such as powerplants or oil refineries. It is important to establish the variables that may be influencing the location of these polluting sites. Since we are concerned about environmental equity, race variables are certainly obvious variables. The race variables to be used in this analysis are Asian, Black, Hispanic Origin and Native American. As explained below, the number of polluting sites in a particular zip code area may also be explained by several socioeconomic variables as well such as education levels, income, unemployment rate, and percentage of residents that are renters.

Organizing the variables into equation form for estimation yields the following equation:

$$\text{Sitedensity} = B_0 + B_1\text{Asian} + B_2\text{Black} + B_3\text{Hispanic} + B_4\text{NativeAm} + B_5\text{College} + B_6\text{Homevalue} + B_7\text{Income} + B_8\text{Population} + B_9\text{Renting} + B_{10}\text{Unemployment Rate}$$

The dependent variable (Sitedensity) is the number of newly permitted stationary sources of air pollution in a given zip code divided by the square miles of the zip code area. The number of sources is measured by the number of new sites permitted for construction in years within two years of when the independent variables were collected by the Census Bureau (e.g., 1988 - 1992). This is critical because, by using the characteristics of a community at the time the decision to build a polluting site was made, information regarding the siting process is uncovered.

There are three types of independent variables. The race variables are the key variables for testing whether ethnicity of the population at the time of siting has a significant influence on the density of newly permitted sites in an area. Variables such as College and Income are included for two reasons. First, including College and Income allows us to control for demographic influences when assessing the significance of the minority variables. If environmental racism is operating, then site density should be high in middle income minority communities as well poor minority communities. Second, inclusion of demographic variables allow us to test whether construction of new point sources of air pollution are being targeted at low income areas in general, irrespective of the racial composition of these areas. Significance of these variables would suggest "classism" rather than "racism" may be occurring.

%Rent, or the percentage of renter occupied units, is included in the model as a proxy for two influences. First, a higher percentage of renters means a lower percentage of homeowners. Since owner-occupied housing is most individuals' single largest investment, there is often strong homeowner opposition to polluting sites likely to drive their property values down. In contrast, renters tend to be more mobile, since they often have the ability to move on a month's notice, or at most have a one year lease. Renters' frequent mobility provides less attachment to a particular community. Thus a higher percentage of renters in an area would lead us to expect a higher site density.

Home value (Homevalue) would be expected to have a negative effect, as this variable serves as a proxy for land values in the area. Higher land values would make an area less profitable for industrial sites. Unemployment rate is expected to be positive because areas with high unemployment may welcome, rather than oppose, new industrial projects. If an area is receptive to new industrial projects (even if it is polluting industry) it may offer incentives for firms to locate there and would certainly not require expensive mitigation actions of new firms.

With this model we can distinguish between the race variables and other competing explanations of the high correlation of pollution sources and minority population. This equation is in essence a simple model of stationary source siting location decisions within a given metropolitan area. It does not explain other factors such as why that many plants are being built (such as the demand for electricity or oil refining).

As will be described in Chapter 6, collection of data from several sources allows an analyst to statistically estimate the magnitude of each independent explanatory variable on the dependent variable (site density). That is, with data and least squares regression we can compute specific values of the B's (slope coefficients) in equation 1. From the estimated slope coefficients we can tell: (a) whether a particular explanatory variable has a systematic influence on the dependent variable (i.e., is there a statistically significant relationship), and (b) the relative magnitude or contribution of each independent variable to the dependent variable. Thus equation 1 allows us to go beyond the level of correlation saying there is a relationship to explain what factors lead to this relationship. This simple model helps us to analyze whether policies explicitly focusing on racial characteristics in plant siting decisions are the right policy variables, or whether other policy variables such as renters would be more effective to address the concern. You will see the results of this analysis in Chapter 6.

Chapter 6

Statistical Analysis Using Regression

This chapter introduces the concept of a statistical analysis technique called multiple regression. Multiple regression provides two very powerful roles for policy analysis: explanation and prediction. Explanation, typically in the form of testing a hypothesis, permits an analyst to identify what factors influence other variables – for instance, is ethnicity related to the likelihood that someone lives near a noxious environmental facility? Will a gasoline tax have an identifiable effect on people's driving behavior? Secondly, as we have stressed in previous chapters, an important task in policy analysis is to predict the effects of each alternative prior to choosing one. A regression, by explaining what factors influence other variables, can be used to predict the effects of changing those factors. If a tax of $0.50 per gallon is placed on gasoline, a regression equation will inform the analyst how much people will change their driving.

Regression analysis is widely used in both the sciences and the social sciences to explain and predict. This chapter illustrates how regression can be used in policy analysis. It provides an overview of the technique, guidelines for how to develop a regression and how to interpret its results, and ways to measure the power of the regression results. Our intent is to show you how to interpret and to use a regression, and perhaps even give you ideas about how to perform one. There is far more to learn about regression than this chapter can possibly explain, but we hope to provide you with some understanding of this very powerful tool.

1. WHAT IS A REGRESSION?

Often, a variable of interest to a policy analyst – for instance, the quantity of gasoline a family purchases, or proximity of someone's home to a hazardous waste site – is influenced by a number of factors, such as the kind of car a family drives and where the family lives relative to jobs, or race and income. A regression is a statistical method that uses *observations* – data on the relevant variables – to identify the contribution of multiple *independent*, or *exogenous*, variables to a *dependent*, or *endogenous*, variable. The independent variables predict the dependent variable: that is, if you are given the independent variables and your regression model, you can come up with a good estimate of the dependent variable.

For instance, suppose that you wanted to develop a model to predict the height of a person as a fully grown adult. That height is the dependent variable. One way to estimate a random person's height is to take the average height of the population and use that as a prediction. While that method works well on average, it does not tell you much about any particular individual. You would like to use characteristics of an individual to predict that person's height. After some research, you identify several factors that influence an individual's height as an adult. Those might include the heights of that person's parents and that person's gender, along with nutrition as a youth. Your model then becomes:

(6-1) Height of an adult = Function of (Height of father, Height of mother, Gender, Nutrition)

Now you collect data, not only on the height of everyone in your sample, but also on their parents' heights, their gender, and their nutrition as a child. You can graph each independent variable against your dependent variable; you will probably get a scattering of points, but perhaps with an identifiable trend to them. You may believe that each of these independent variables makes some contribution in a *linear* fashion to determining height, though you don't know the relative contributions of each factor. The magnitude of the effect of each independent variable on the dependent variable is measured by the coefficients or Bs in the equation below. The equation is:

(6-2) Height of an adult = B_0 + B_1*Height of father + B_2*Height of mother + B_3* Gender + B_4*Nutrition

The purpose of running the regression is to estimate the most likely numerical relationship between your independent variables and your dependent variable, i.e., the numerical values of the Bs. A regression takes all the independent variables into account and deciphers the contribution of each factor individually to height. Running the regression tells you the *coefficients*, the change in the dependent variable associated with a change in each independent variable. For instance, once you have run your data through a multiple regression computer package, the resulting equation might be:

(6-3) Height of an adult, in inches = 2 + 0.5*(Height of father) + 0.46*(Height of mother) + 2*(1 if male, 0 if female) – 2*(1 if poor nutrition, 0 if good nutrition),

where the first 2 is referred to as the intercept (since it is not multiplied by an independent variable), 0.5 is the coefficient on Height of father, 0.46 is the coefficient on Height of mother, etc. Now, if you know how tall a person's parents are, whether the person is male or female, and the quality of the person's diet while growing up, you can use this regression equation to predict that person's height. For instance, if a woman's father was 70 inches tall, her mother 64 inches tall, and she had a good diet growing up, then plugging these values into this equation provides the prediction that she would be 66.94 inches tall, while her brother (with the same nutrition) would be 2 inches taller (because the coefficient on Male adds 2 inches to height).

If you go back to the original data, you can now predict, for any person in your data set, how tall you expect that person to be, just as we did with the woman and her brother in the previous paragraph. When you actually find the woman and her brother, though, you discover that in fact she is 65.5 inches tall (1.44 inches less than predicted), and her brother is 70 inches tall (1.06 inches taller than predicted). Does that mean that your regression is invalid? Remember that your regression is a way of *estimating* the relationship between the independent and dependent variables when you don't know the underlying truth. You should not expect to get exact answers (except in some unusual cases where there is a true physical relationship between variables, such as with some constants in physics). Instead, you hope to find *statistically significant* relationships between the independent and dependent variables. Statistical significance refers to a high probability (typically, more than 90 or 95 percent) that a coefficient is different than zero. If a coefficient is significantly different than zero, then the variable with which that coefficient is associated does influence the dependent variable; if there is no statistically significant relationship, then the coefficient might be zero, and changing the independent variable might not affect the dependent variable.

How can you determine statistical significance? The same computer package that gave you your coefficients probably gave you, as well, the *standard error* and the *t-statistics* for your coefficients. The standard errors provide an estimate of the variation around the estimate of the coefficient – that is, whether the range of error around the coefficient estimate is small or large. The t-statistics (calculated as the coefficient divided by the standard error) are a way of measuring whether the coefficient is statistically significantly different from zero. A t-statistic of about 1.7 or higher (in absolute value) suggests statistical significance; below that level, the estimate is unreliable, and the independent variable should usually be treated as having a zero effect on the dependent variable.

For instance, suppose the results from your regression, above, gave the following t-statistics along with each coefficient:

Table 6-1. Possible Regression Results

Independent Variable	Coefficient	Standard Error	T-Statistic
Intercept	2	1.7	1.18
Height of father	0.5	0.33	1.51
Height of mother	0.46	0.2	2.3
Gender	2	0.9	2.22
Nutrition	-2	1.1	-1.82

Coefficients that are clearly statistically significant from zero include height of mother, gender, and nutrition, since, in absolute value, they are all above 1.7. Height of father approaches statistical significance but does not quite achieve that goal, and the intercept is also not statistically significant. In other words, this regression has found significant relationships for three of the five variables, but it has not demonstrated a relationship between height of father and an adult's height. This issue of statistical significance is discussed further below.

Note that the causality runs from the independent variables to the dependent variable. A child does not cause her parent's height, nor does a child's height cause its gender or diet. In a regression model, careful thought needs to be given to what is dependent on other factors. The causal relationship should only go one way: the independent variables influence the dependent variable, but a dependent variable should not influence the independent variables. If no variable is dependent on another variable, then there is no predictability: the occurrence of sunspots, for instance, is unlikely to have any influence at all on someone's height, or vice versa. If two variables are likely to influence each other – for instance, a person's height is likely to influence her weight, and weight may influence height through nutrition – then special thought needs to be given to what exactly is being studied. Perhaps some independent variables predict each separately, or perhaps each affects the other. More advanced statistical methods may be necessary in such cases.

Learning how to calculate the effects of changes in the independent variables in a regression will enable you to provide useful predictions for policy analysis. For example, the demand curve for beach recreation is influenced by the quality of water at a beach, price (in distance and money) of travel to the beach, a person's income, and availability of substitute vacation spots. An example of such a regression is:

(6-4) Trips per year to beach A = 4 + 0.2*(Water quality index, on a 0-10 scale) – 0.05*(Cost of traveling to the beach, parking, and admission) + 0.04*(Income, in thousands) – 0.3*(Number of alternative beaches in a 10-mile radius)

As will be discussed in Chapter 7, having a demand curve for recreation permits calculation of the benefits associated with that recreational use. The benefits (or costs) of changing water quality can be estimated using this regression equation, by changing the water quality variable and re-calculating the benefits of the recreational resource.

Among the independent variables, it is useful to distinguish between policy variables, which can be controlled by policy makers, and state-of-nature variables, which cannot be influenced by a decision-maker. For example, suppose that the number of bus trips a person takes per day is estimated as:

(6-5) Bus Trips = 2.2 – 0.06*(Income, in thousands) + 0.05(Age) + 0.05(Distance to work, miles) – 0.5(Bus fare, in dollars) + 0.4(Price of gasoline, in dollars).

Income, average age of the population, and distance a person lives from work are often taken as state-of-nature variables in the short run. It is hard to imagine a transportation policy that can have any effect on the age distribution of the population. Yet the age distribution of the population influences the demand for mass transit. In contrast, bus fare and price of gasoline can be influenced by a policy-maker through tax or subsidy measures.

2. MULTIPLE REGRESSION MODELS AND CONTOUR MAPS

Consider again the regression equation for beach recreation:

(6-6) Trips per year to beach A = 4 + 0.2*(Water quality index, on a 0-10 scale) – 0.05*(Cost of traveling to the beach, parking, and admission) + 0.04*(Income, in thousands) – 0.3*(Number of alternative beaches in a 10-mile radius).

The simplest demand relationship is the relationship between two variables -- price (cost in this equation) and quantity, holding the other determinants of demand constant. This is an over-simplification, for our choice of how many units to consume depends on these additional variables. When economic problems involve more than two variables, as in this equation, a two-dimensional graphic illustration is not enough. Here we want to keep track simultaneously of 4 independent variables. Baumol and Blinder suggest that we think of the shifts in demand resulting from the introduction of additional variables as analogous to the contour information presented on detailed maps of mountain areas in the United States.

The U.S. Geological Survey has developed a well-known device for collapsing three dimensions into two, namely a "contour map." On contour lines or "rings," we find a number indicating the height above sea level at that particular spot on the mountain. All points on any particular contour line represent geographic locations that are at the same height above sea level. Thus, unlike the more usual sort of map, which gives only latitudes and longitudes, the contour map exhibits three pieces of information about each point: latitude, longitude, and altitude.

Figure 6-1 looks more like the contour maps you will find in graphing a multiple regression. It shows how some third variable I, which might be consumer income, affects the relationship between variable Price and the quantity variable Q. Just like a contour map, any point on the diagram conveys three pieces of data. In this contour map, all points on a given contour line represent the different combinations of direct costs or price paid and trips to the beach by representative individuals receiving a given level of income. For example, all points on demand curve A with I = $25,000 (and with water quality fixed at 7 and number of alternative beaches at 3) represent the demand for beach recreation at various levels of price. When this information is put into the above equation, it becomes:

(6-7) Trips per year to beach A = 4 + 0.2(7) – 0.05(Cost of traveling to the beach, parking, and admission) + 0.04(25) – 0.3(3):

= 5.5 – 0.05(Cost of traveling to the beach, parking, and admission).

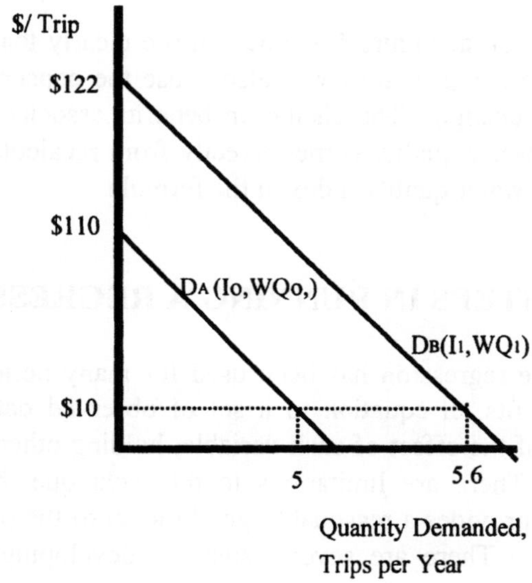

Figure 6-1. Demand Shifts as Contour Lines

In our graph, a change in the cost of traveling to the beach produces a "movement along" a fixed demand curve. By contrast, a change in any other variable that influences quantity demanded – in this example, income, water quality, or number of alternative beaches -- will produce a "shift" of the demand curve. For instance, if water quality improves to 10, the equation becomes:

(6-8) Trips per year to beach A = 4 + 0.2(10) – 0.05(Cost of traveling to the beach, parking, and admission) + 0.04(25) – 0.3(3)

= 6.1 – 0.05(Cost of traveling to the beach, parking, and admission).

Before the water quality improvement, a $10 cost of travel to the beach would result in 5.5-(0.05)*10 = 5 trips to the beach; with improved water quality but no change in travel cost, the number of trips will increase to 6.1 – (0.05)*10 = 5.6 trips. In the price-quantity world of a demand diagram, it is as though the intercept of the demand curve (and sometimes, for some equations, even the slope!) shifted in response to changes in these other factors.

If you look at Figure 6-1, you can see clearly that any event that causes the demand curve to shift will also cause the consumer surplus estimate of benefits to change. The change in benefits associated with, for instance, a change in water quality comes directly from recalculating consumer surplus with a new water quality index in the formula.

3. STEPS IN RUNNING A REGRESSION

Multiple regression has been used for many policy relevant phenomena because it fits an equation to a set of observed data, providing statistical estimates of the effect of each variable, holding other variables in the model constant. There are limitations to the technique, but multiple regression frequently provides a reasonably good model of the phenomena or behaviors of interest. There are several steps in developing a multiple regression model.

3.1 Specify the Variables in the Policy Model

The first step in multiple regression analysis is to specify the variables that are expected to influence the dependent variable. In a demand function for a good such as electricity, commuting trips, gasoline, etc., the dependent variable to be explained is the quantity demanded (Q). The list of independent variables that influence demand always includes the price (P) of a good and generally includes such factors as consumer income (I), the price and availability of substitutes (PSubs), other socioeconomic variables such as age (A) and education, and sometimes an indicator of the quality of the good. If aggregate demand for a city or state is estimated, then the size of the population of the consuming group (K) will need to be included. Alternatively, if the regression uses data on individual consumers, then measures of individual taste or preferences (D) obtained from survey data may be included. Of course capital stock variables of the household may be important when predicting the demand for gasoline, water or electricity. For these three goods, how many cars the household owns, the size of the yard, and whether the house has central air conditioning would help explain variations in the household's demand for gasoline, water, and electricity, respectively.

Note that economic theory provides information on what variables are relevant to the demand equation. If you review an economics text on factors that influence consumer demand, you will see mention of the price of the good, the price of related goods (complements and substitutes), income, and tastes and preferences (the socio-economic variables try to capture these effects). If, in contrast, you were running a regression on the effects on cancer rates of exposure to a pollutant, you would draw on medical theory to identify the other factors likely to influence cancer rates, such as smoking behavior, age, and genetic predisposition. It is highly important to think about your regression before you collect your data and run your analysis. Better modeling in advance can lead to more satisfactory regression results through inclusion of all the appropriate variables.

We will illustrate a regression model with a simple demand function. The individual demand function for a good (x) is a statement of the relationship between the quantity demanded and all of the factors that affect this demand. Written in general form, the demand function may be expressed as:

(6-9) $Q_X = f(P_X, \text{Income, Tastes, PSubs, Age, etc.})$

If important determinants of the dependent variable are inadvertently omitted from the regression model, the equation may not accurately predict the effect of a potential policy on the dependent variable. For instance, in a demand curve for electricity, the size of a house affects electricity consumption as well as whether the house has central air conditioning (CAC). Suppose that larger houses are more likely to have central air; both lead to greater electricity consumption. Suppose, too, that you don't have data on house size, only on whether a house has CAC, and you omit house size from the regression. Then, in your regression equation, the coefficient on CAC will include some of the effect of a larger house on electricity demand and thus overstate the effect of CAC. If you were to try to predict the effect on electricity demand of converting a small house from no air conditioning to CAC, your prediction would include, in part, house size and thus overstate your electricity demand from switching.

3.2 Choosing a Functional Form: Linear or Non-Linear?

Once the variables have been chosen, the next step is to specify the form of the equation, or the manner in which the independent variables are assumed to interact to determine the level of the dependent variable. Multiple regression is often termed linear regression, a term that makes people think that a linear relationship between the dependent and independent variables is the only one possible. In fact, the "linear" in linear regression only refers to the coefficients, not to the dependent or independent variables. While:

(6-10) $Y = B_0 + B_1X_1 + B_2 X_2$

is obviously a linear regression, so are the following:

(6-11) $Y = B_0 + B_1X_1 + B_2X_1^2 + B_3X_2 + B_4X_2^2 + B_5X_1X_2$

(6-12) $Y = B_0 + B_1\ln(X_1) + B_2\ln(X_2)$ (where ln refers to the natural logarithm)

(6-13) $Y = B_0 + B_1X_1 + B_2X_1^{1/2} + B_3X_1^{1/4} + B_4X_1^{1/8} + B_5[\ln(X_2)]^2$.

In each case, even though the relationship between X_1 and Y or between X_2 and Y is more complicated than the simple linear one above, the coefficients are still linear and can be estimated using ordinary least squares, the classical linear regression method.

If you were to graph the relationship between the dependent variable and an independent variable, sometimes it makes sense for the relationship to be a straight line. For instance, if you double the use of an input to production, you are likely to double the cost associated with that input. In other cases, though, a straight line might not make sense. For instance, if you were graphing the relationship between an input to production (such as fertilizer) and output (such as crop yield), the theory of diminishing marginal product would argue that doubling fertilizer would not be expected to double yield; yield would increase more slowly. In this case, a linear relationship between yield and fertilizer would not be appropriate, and a functional form that allowed for a nonlinear relationship would be more appropriate.

To illustrate, let us assume that we are analyzing the demand for commuting trips to work. If the demand function has been specified as a linear relationship, each independent variable enters into the equation as itself, and only itself:

(6-14) $Q = Bo + B_1Price + B_2Income + B_3Tastes + B_4PSubs + B_5Age$

Linear demand functions, such as this, have great appeal in empirical work for two reasons. First, the slope or regression coefficients (Bo to B_5)

provide a direct measure of the marginal relationships in the demand function. That is, they indicate the change in quantity demanded caused by a one-unit change in each of the related variables. For example, if $B_1 = -2$, quantity demanded will decline by 2 units per time period with each $1 increase in the price. Second, sometimes demand relationships are in fact approximately linear over the range in which decisions are made. The demand curve derived from a demand function such as this is a straight line, as shown in Figure 6-2.

If a linear relationship between the dependent and independent variables is not appropriate, there are numerous other functional forms that can reflect nonlinear relationships, as illustrated above. Figure 6-2 illustrates some of the most common functional forms used in regressions, including quadratic, double-log, and semi-log.

The quadratic equation includes, in addition to the linear variables, squared terms for one or more of the independent variables:

$$(6\text{-}15)\ Q = Bo + B_1\text{Price} + B_2(\text{Price})^2 + B_3I + B_4I^2\text{, etc.}$$

In this form, a one-unit change in price now has a more complicated effect on quantity: dQ/dP, the slope of the demand curve, is now $B_1 + 2B_2\text{Price}$. The relationship between price and quantity (or quantity and income, for that matter) is no longer constrained to be a straight line. If B_1 is negative, the first-order effect is that the demand curve slopes down; however, depending on the sign of B_2 (that is, whether it is positive or negative), the demand curve may become steeper or more shallow as price increases. Because the quadratic form is shaped like a parabola, with either a maximum or minimum, in some cases this shape may result in nonsensical results in some range of the policy analysis. For instance, as price goes up, quantity may decrease over one range but increase over another. While this shape might make sense for other kinds of regressions (for instance, in cases of extreme diminishing marginal product), it may make less sense for a demand curve. Still, if the analyst is aware that the resulting equation should only be used in a limited range of the independent variable, the quadratic form can be useful in many cases.

The double-logarithmic equation takes the natural (i.e., to the base e) logarithmic form on both sides of the equation:

$$(6\text{-}16)\ \ln Q = Bo + B_1(\ln \text{Price}) + B_2(\ln I) \ldots\ldots$$

This expression is equivalent to the so-called Cobb-Douglas functional form,

(6-17) $Q = e^{B0}(\text{Price})^{B1}(\text{Income})^{B2}\ldots$

(where **e** refers to the exponential, sometimes called the antilog, of 1, 2.71828. . .).

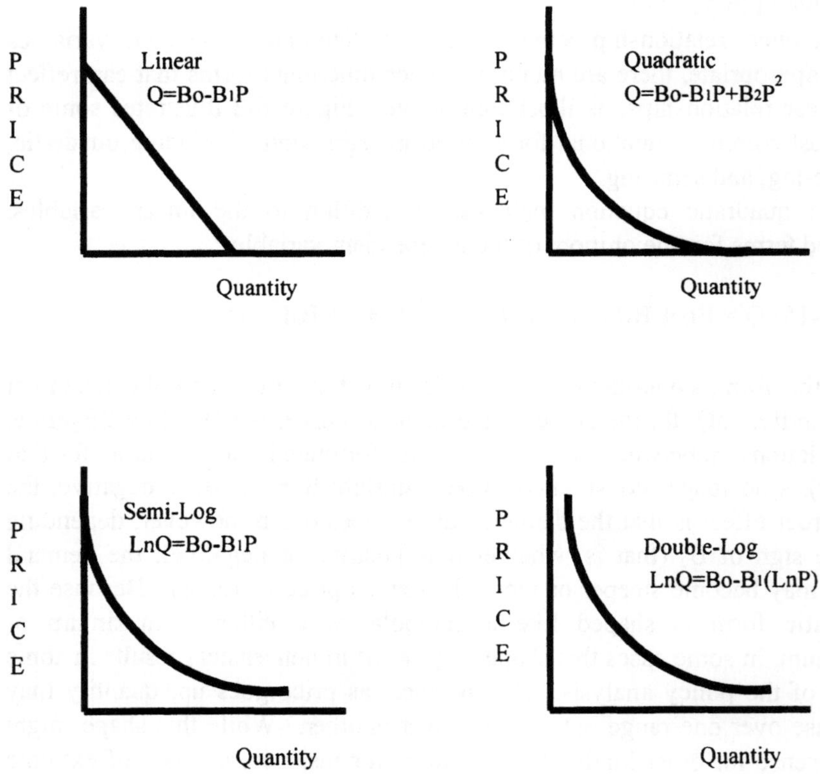

Figure 6-2 Common Functional Forms used in Regression Analyses

Depending on the sign and size of the Bs (the coefficients), the relationship between Q and an independent variable could be increasing or decreasing at either an increasing or a decreasing rate; thus, again, a curved relationship is possible between the dependent and independent variables. Here the graph of the relationship between the dependent and an independent variable does not have a maximum or minimum; thus, for some data sets it may be more desirable than the linear or quadratic. On the other hand, the double log may not be desirable for a demand curve, because there is no finite "choke" price that will lead to quantity demanded going to zero (a likely result of very high prices for most goods).

The unique advantage of the double log form is the ease with which the regression coefficients can be compared. It is the only functional form in which the regression coefficients are the elasticities, the percent change in the dependent variable associated with a one percent change in each of the independent variables. For instance, if $B_1 = -0.5$ and $B_2 = 1$ in the above equation, then a 10% increase in price leads to a 5% decrease in quantity purchased (5%/(-10%) = -0.5), while a 10% increase in income leads to a 10% increase in quantity purchased. Because the coefficients are elasticities, their magnitudes can be directly compared to provide an accurate indication of which independent variable has the largest effect on the dependent variable. This cannot be done with a linear functional form because the magnitude of the B coefficients depend on the scale of the units of measurement for each independent variables.

In the semi-logarithmic form, either the dependent variable or one or more independent variables may be logged. Thus, the semi-logarithmic equation can be written as:

$$(6\text{-}17)\ \ln(Q) = Bo - B_1 Price + B_2 I \ .....$$

or as

$$(6\text{-}18)\ Q = Bo - B_1 \ln(Price) + B_2 \ln(Income) \ .....$$

In many recreation demand studies, applications of the semilog form frequently take the first form, converting number of trips to a natural logarithm and leaving all independent variables in their linear form. This form has a unique advantage in estimating the average consumer surplus per unit Q as simply one divided by the regression coefficient for price.

A large range of functional forms is possible for any regression. Identifying a form that fits the data well is often the reason for running many regressions on the same data set. As suggested above, theory can be used to suggest characteristics of the functional form. For instance, it is likely that a production function (a relationship between input and output) will display diminishing marginal product – a one-unit change in input use will lead to a decreasing amount of additional output – and thus one of the non-linear forms suggested above would be appropriate. In some production functions, it might be true that the amount of one input can influence the marginal productivity of another input. For instance, ground-level ozone, a pollutant, is produced by a chemical reaction between hydrocarbons (HC) and nitrogen oxides (NO_x) in the presence of sunlight. The amount of ozone produced when NO_x is increased depends on how much HC is present. If you were to run the regression:

(6-19) $\text{Ozone} = B_0 + B_1HC + B_2NO_x + B_3(HC*NO_x)$,

then the effect of one more unit of NO_x on ozone production would be:

(6-20) $d(\text{Ozone})/d(NO_x) = B_2 + B_3HC$

now depends on the level of hydrocarbon, due to the use of this *interaction* term (B_3), as theory predicts.

In other cases, theory may not suggest what form the equation should take. In these cases, you should let the data suggest the functional form and the form that provides the best possible fit of the data. Scatter diagrams of the relationship between the dependent variable and each of the independent variables may suggest which form reflects the true relationships between them. In practice, several different forms may be tested, and the one that best fits the data (using some of the statistical tests described below) should be selected as being most likely to reflect the true relationship.

3.3 Collect the Data

The next step in regression analysis is to obtain observations on the variables. While this step is conceptually easy, in practice it can be a major hurdle. Sometimes your data come from different sources and thus may not be immediately comparable – for instance, you may have daily prices for gasoline in an area, but only weekly data on the amount consumed. In these cases, you will need to decide how to convert one set of data to the other units. You also need to consider seriously the effects of not being able to find the right kinds of data. For instance, if you cannot get any data on the cost of public transportation in your area, your regression results may be *biased*, meaning that the other regression results will be distorted through omission of this variable. Additionally, you need to have data where there is variation in the individual observations. For instance, if you do have the price of public transportation, but it was constant during the period of your study, then there is no way to identify how changes in public transportation prices are going to influence the demand for gasoline.

As these comments suggest, you need to think about what results you would like to present when you decide how to collect your data. If you want to look at the influence of public transportation prices, then you should probably consider collecting *cross-sectional* data – data from a number of different places at the same time – since different places are likely to have different transportation prices. On the other hand, if you want to look at how drivers in one area respond to changes in the price of gasoline, you should focus on getting data from that area – most likely, *time series* data, that provide changes over time -- and not worry about changes in the price of public transportation. In other words, your data should reflect the nature of the problem you are analyzing.

3.4 Running Your Regression

Regression equations are typically calculated -- that is, the coefficients, or Bs, are estimated -- by the statistical method of least squares. Consider, for now, a regression with only one independent variable and an intercept:

(6-21) $Y = B_0 + B_1X_1$.

If put your data on a graph, you will see a scatter plot of points that (you hope) bear some resemblance to a line. The job of the regression is to draw a line through the data that is as "close" as possible to every single point. Now, draw a line from each data point to the regression line. You'd like to minimize the size of all these line segments. If you just add up the lengths of those segments, though, then the way to minimize the sum of those segments is to make each segment negative (and as negative as possible). Obviously, that approach violates the spirit of what you want to do.

The method of Ordinary Least Squares (OLS), or linear regression, gets around this problem by minimizing the sum of the *squares* of the difference between each of the actual data points and the estimated line. Squaring the deviation of each data point from the fitted line, or error term, produces a sum of a set of positive numbers. The line that minimizes this sum most accurately depicts the relationship between the dependent variable and each independent variable. The procedure is presented graphically in Figure 6-3. Here each point represents the price and quantity of gasoline used by an individual in our sample. It shows the horizontal deviation of each data point from the fitted line.

When more than one independent variable is included in the regression, the theory of how the regression works is the same: the sum of squared error terms is minimized for each independent variable.

Most spreadsheet programs and nearly all statistics packages have regression programs. All you need to know is how to put the data into the program and, most importantly, how to interpret the output. Accordingly, we shall concentrate on setting up multiple regression problems for computer solution and interpreting the output, rather than dwelling on the mathematical process itself.

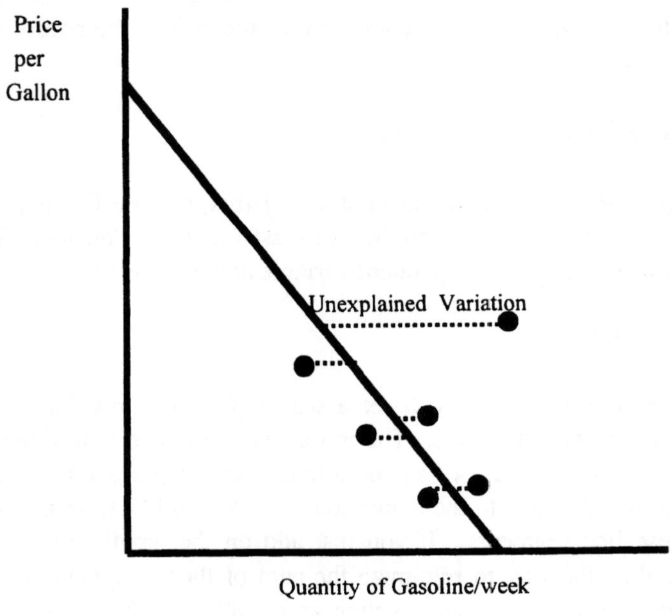

Figure 6-3 Scatter Plot of Data Points, and Regression Line

3.5 Interpreting Regression Coefficients

Learning how to interpret the many regression statistics which are typically provided by computer-based regressions can teach you a great deal about the problem you are studying, by telling you what influences your dependent variable. It can also enable you to make predictions of the effects of alternative policies. Getting these results from a regression begins with learning how to interpret the magnitudes of the regression coefficients.

The key information provided by statistical model is the regression coefficient for each of the variables, such as B_0 through B_5 in Equation 6-14. Each coefficient indicates the how much a change in each independent variable changes the dependent variable, holding constant the effects of all other variables included in the model. For example, assume that the computer printout of a statistical demand function for automobile commuting trips to work each month is:

(6-22) Q = 21 -1.1Price + 0.0001Income + 0.54PSubs -.088Age

(t statistics) (3.06) (-5.03) (6.16) (2.35) (-2.89)

N = 200 Adjusted R^2 = 0.82 F = 126.4
Standard error of the estimate = 0.25.

The values of the regression coefficients provide us with an estimate of the change in number of car trips to work each week associated with a one-unit change in each of these independent variables, holding constant the effect of all other variables. For example, Equation 6-22 indicates that individual monthly demand for auto trips falls by 1.1 trips with each $1 increase in the price. Thus a policy that increased gasoline costs by enough to raise the average work trip cost by $1, or increased the price of parking by $1, would lead to 1.1 fewer trips, on average. The coefficient for income indicates that, if nothing else in the equation changes, a $1,000 increase in income will cause the number of trips to increase by 1 (= $1,000 * 0.001) per month. The coefficient for the price of substitute transportation methods shows that demand for auto trips to work increases by 0.54 of a trip with each $1 increase in the average price of substitute ways to travel to work (e.g., bus fares).

The constant or intercept term has a value of 21 in Equation 6-22. This term represents the demand for auto commute trips to work that is not explained by the independent variables included in the equation. It can be viewed as the expected number of car trips when all the other independent variables are assigned values of zero. It may have no meaning if the observed values for the independent variables are never around zero. It is hazardous to generalize about the relationship between the independent variables and dependent variable beyond the range of observed values, since you don't know how well your regression equation works for those ranges.

3.6 Evaluating Regression Results Using Other Statistical Information

As is probably clear by now, you have a number of decisions in running a regression – what variables to include, what functional form to use, and what sources of data to employ, for instance. How do you decide whether you've done a good job? In part, the regression itself can tell you. Some of the information provided by the statistical package (and any package that runs a regression will provide you with at least some of this information) can give you some guidance in how well your regression worked.

Look again at Equation 6-22, above. In addition to the regression equation itself, the information provided includes t-statistics for each coefficient in the regression, the R^2 for the equation, an F statistic, and the standard error of the estimate. Each of these can tell you about the quality of your regression.

3.6.1 The T-Statistics

How can you determine whether a particular independent variable has an influence on your dependent variable? If the variable has no effect, then its coefficient should be zero: a change in the independent variable leads to no change in the dependent variable. Of course, just due to chance the regression equation might find some effect other than zero. To determine whether a non-zero coefficient is in fact different from zero, you can use the t-statistics provided by the regression equation. These numbers, in parentheses below the regression coefficients in Equation 6-22, measure how many standard errors of the coefficient lie between that coefficient and zero. The t-statistic is calculated by the computer program by dividing the coefficient by the "standard error of the coefficient." The standard error of the coefficient is like the standard deviation; it provides a measure of the variability of the coefficient. If the standard error is quite small relative to the magnitude of the coefficient, then we can place more confidence in our conclusion that this particular variable has a systematic (i.e., non-zero) effect on the dependent variable. A frequent test of whether the variable is significant is whether each regression coefficient is at least twice its standard error. This is equivalent, in most cases, to observing a "t-value" of approximately 2 for a regression coefficient. If this is the case, there is less than a 5 percent chance that the independent variable is not significantly related to the dependent variable. An equivalent way of thinking of this is in terms of the confidence interval around the coefficient. If a variable is statistically significant at the 5% level, then the 95% confidence interval around the coefficient does not include zero.

In our example above, all of the variables are significantly different from zero. However, we often learn as much from what variables are insignificant as those that are significant. That is, it would be quite valuable when evaluating alternative policies to find out if one of the policy variables actually has no systematic effect on the environmental parameter of interest. In this case, we can rule out that policy variable, as it would be a totally ineffective instrument to achieve the environmental policy. For example, if we find that that the presence of a carpool lane in other cities has no detectable effect on vehicle miles travelled, this would be quite useful in suggesting that policy instrument may not work in our city.

The rule of thumb for interpreting a t-statistic is that a value of around 2 or higher is statistically significant, while a t-statistic below about 1.5 is not significantly different from zero at commonly used alpha levels of 10% or higher. Like any rule of thumb, though, it needs to be viewed cautiously. In fact, it is possible that a variable does in fact have a significant effect, even if your t-statistic suggests that it does not. One common reason for insignificance is a small data set, and the associated large variance. Identifying statistical significance is in part dependent on the number of observations: with more data, it is easier to identify a statistical effect (or to be confident of no statistical effect). For example, think back to estimating the average height of the people in your city. If you only have the heights of a few people – say, the people in your family – you are not likely to be very confident of your results. If you sampled a thousand people in your city, you would be much more confident that your estimate of the average height in your city was good. Similarly, for any statistical analysis, your results will be more reliable with more data.

Another possible reason for insignificance when there really is an effect is *multicollinearity*, a situation where two independent variables are closely related to each other. The t-statistics identify the effect of each independent variable separately. If two independent variables are closely correlated -- for instance, the price of gasoline and the price of motor oil, both petroleum products – then the effect of either variable separately from the other is small, though the effect of these variables together is large. The two variables are, in a way, sharing explanatory power between them; though neither is significant on its own, they might be significant together. It is possible to test for the joint significance of variables using an F test (described below). It is also possible, if two variables are highly correlated (calculated as a correlation coefficient), to drop one from the regression equation and rely on the second, if the separate effects are not considered important. There is also no harm to leaving both in the regression equation, even though the variables are not statistically significant. A statistics or econometrics book can provide more ideas on how to cope with multicollinearity in a regression.

In sum, the t-statistics are your first suggestion whether an independent variable influences your dependent variable. They provide an important explanatory role in regressions. Significance of a variable in a regression provides strong evidence that that variable in fact influences the dependent variable and can, in and of itself, provide important policy information. For instance, a great deal of research on environmental justice hinges on whether race and income are statistically significantly related to the location of a hazardous waste site or other environmental bad. Being able to interpret t-statistics enables you to understand and critique others' analysis as well as to understand your own regression results.

3.6.2 R^2 as a Measure of Goodness of Fit

The coefficient of determination, identified by the symbol R^2, is defined as the proportion (or percent) of the variation in the dependent variable that is explained by the full set of independent variables included in the equation. In other words, R^2 is an approximate indicator of how well you've done in explaining your dependent variable. In equation (6-22), it indicates that the regression explains 82 percent of the variation in the dependent variable. R^2 can have a value ranging from 0, indicating the equation explains none of the variation in the dependent variable, to 1.0, indicating that all of the variation has been explained by the independent variables.

Regression models will seldom have an R^2 equal to either 0 or 1.0. In empirical demand estimation R^2 values of 0.50, indicating that 50 percent of the variation in demand is explained, are quite acceptable. For aggregate or time series regression models, R^2s of 0.80 or even slightly higher are sometimes obtainable. However, for regression models based on individual consumer or producer behavior, you must be satisfied with considerably less explanation of variation in demand. When the coefficient of determination is very low -- say, in the range of 0.05 to 0.15--it is an indication that the equation may be inadequate for explaining the effect of the policy on human behavior. The most general cause of this problem is the omission of an important variable or variables from the equation or model. Thus, the R^2 can be used as a feedback mechanism to both the analyst and the decision maker regarding the need for additional variables and associated data.

Because R^2s of 1 are virtually impossible to attain, and because the value of R^2 for an equation varies somewhat with the nature of the data with which you're working, having a high or low R^2 in and of itself is not highly useful. Where it can be especially useful is in selecting among different regression models that you may run. For instance, suppose you run both the following equations, using the same set of data:

$$(6\text{-}23)\ Q = B_0 + B_1\text{Price} + B_2\text{Income}$$

$$(6\text{-}24)\ Q = B_0 + B_1\ln(\text{Price}) + B_2\ln(\text{Income})$$

If one has a much higher value for R^2 than the other, then the regression with a higher R^2 explains more of the variation in Q and is thus a better fit for the data.

Two additional points should be made with respect to the coefficient of determination, R^2. First, most computer programs print out two coefficients of determination. The first, referred to as R^2, is a straight calculation of the amount of variation explained by the independent variables; the second, adjusted R^2, is adjusted (in an imprecise way) for the number of variables included. Adding an independent variable to a regression can't hurt your regression results: if it is unrelated, it will leave the rest of your regression alone. If it is related, of course you want it included. The harder cases are where adding a variable has an effect, but you're not sure if it's real or just by chance. The unadjusted R^2 will never go down, and might (just by chance) go up due to including more variables. The adjusted R^2 adds a penalty for adding a variable; it might not go up when a new variable is added, because the new variable may add something just by chance to the equation. The adjusted R^2 is a more appropriate measure, since it puts a burden of proof on the new variable to justify its inclusion in the regression.

Second, some computer programs show the proportion of the variation in the dependent variable that is explained by each of the independent variables included in the equation, the sum of which equals the adjusted R^2. For example, price may explain 20 percent of the variation, income 10 percent, etc.

3.6.3 The F-Test

The "F-test" or "F-ratio" is used to estimate whether there is a significant relationship between the dependent variable and all the independent variables taken together. Tables are available in any statistical textbook showing the values of F that are exceeded with certain probabilities, such as 0.05 and

0.01, for various degrees of freedom of the dependent and independent variables. The F-ratio of 126.4 indicates the overall equation is highly significant.

As mentioned above in the context of the t-statistic, an F-test can also be used to test whether two or more variables are important to an equation, even if individually they are insignificant. More detail about this approach can be found in Kennedy (1998), p. 56.

3.6.4 Standard Error of the Estimate

Another useful statistic printed out by the computer is the "standard error of the estimate." This statistic provides a measure of the precision of the estimates of the dependent variable from the regression model as a whole. Greater predictive accuracy is associated with smaller standard errors of the estimate. If the errors are normally distributed about the regression equation, there is a 95 percent probability that observations of the dependent variable will lie within the range of two standard errors on either side of the estimate at the mean of the data.

The confidence band concept is illustrated in Figure 6-4. Here we see the least squares regression line and the upper and lower 95 percent confidence limits. Thus, there is only a 5 percent chance that the true estimate lies outside this confidence interval. Note, however, the confidence interval or confidence band is tightest around the center (that is, the mean the observed independent variables) and wider the further one moves from that mean. This suggests that predictions from independent variables that are substantially above or below the mean of the data carry with them less confidence about their accuracy, and the prediction errors are much greater than just two standard errors of the estimate. The interested reader should consult a regression or econometrics book for the details of calculating confidence intervals around regression predictions.

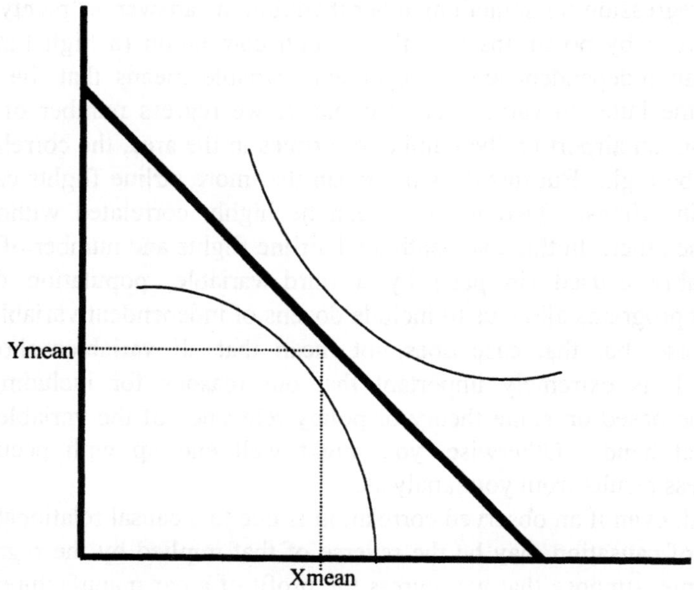

Figure 6-4
Confidence Bands calculated using
Standard Error of Estimate

4. SOME CAUTIONARY NOTES ABOUT REGRESSION ANALYSIS

Regression analysis, like any tool, should not be applied blindly. While, mechanically, any set of variables can be used in a regression against any dependent variable, some regressions make sense, while others are the statistical version of "garbage in, garbage out." As discussed above, thinking about your problem before running a regression can help you avoid various pitfalls. The following discussion is intended to encourage you to think even more about how you set up your regression before you conduct it.

4.1 Potential Regression Problems

So you have a large number of observations on a large number of variables, and you decide to conduct a fishing operation, running regression after regression on different combinations of variables against a range of dependent variables. What can possibly go wrong? If you actually want to use your regression for something other than fun, the answer is "plenty."

First, it is by no means true that a high correlation (a high t-statistic) between an independent and a dependent variable means that the former "causes" the latter to vary. For example, if we regress number of airline flights from an airport on the number of crimes in the area, the correlation is bound to be high. But this does not mean that more airline flights cause an increase in crimes. Two variables can be highly correlated without one causing the other. In this case, both total airline flights and number of crimes are probably caused (in part) by a third variable, population density. Computer programs allow us to include dozens of independent variables with relative ease, but that ease does not mean that all variables should be included. It is extremely important that our reasons for including each variable be based on some theory or policy relevance of the variable to the decision at hand. Otherwise, you might well end up with peculiar or meaningless results from your analysis.

Second, even if an observed correlation is due to a causal relationship, the direction of causation may be the reverse of that implied by the regression. For example, suppose that we regress the profit of a car manufacturer on its advertising expenditures, where profit is the dependent variable to be explained and advertising expenditure the independent variable. If the correlation between these two variables turns out to be high, does this imply that high advertising expenditures produce high profits? Perhaps not. The line of causation could run the other way: High profits could enable management to spend more on advertising. Thus, in interpreting the results of regression studies, it is important to ask whether the line of causation assumed in the studies is correct. Again, you need to draw on theory – in this case, theories of how firms operate – to develop your model.

Third, regressions are sometimes used to forecast values of the dependent variable lying beyond the range of sample data. For example, an examination of observations of road tolls and auto commuter trips may show that the data for the dependent variable, trips, range from 16 to 28, and the independent variable, tolls, ranges from \$0-\$5. If the regression is used to evaluate an extreme policy of very high road tolls (e.g., \$15), both the causal variable and the resulting forecast for the number of trips per month (say, 9 trips) are outside the range over which the original data were collected. This procedure, known as "extrapolation," is dangerous because the available data provide no evidence that the estimated relationship found in your regression holds beyond the range of the sample data. Within your sample, you may make a very good estimate of the effect of changes in an independent variable, but, outside the range of that independent variable, the true relationship between the independent and dependent variable may take a very different form. Just as 2 points are best connected with a straight line but including a third may make a curve more desirable, extrapolating beyond the known data will lead to unreliable results.

Fourth, you should be careful to avoid creating a spurious correlation by dividing or multiplying both the dependent variable and an independent variable by the same quantity. For example, we might want to determine whether a city's auto commuter trips is related to its trips for all purposes. To normalize for differences in population among cities, it may seem sensible to use trips per capita. This procedure may result in these ratios being highly correlated (because the denominator of the ratio is the same for both variables), even if there is little or no relationship between the number of auto commuter trips and trips of all purposes.

Fifth, a regression is based on past data and may not be a good predictor of future values due to changes in the relationships between the variables. For example, past data indicate that age has a negative effect on travel demand. However, better health care and greater physical fitness may cause future consumers not to reduce their travel as fast with age as they did in the past when the data were collected. Thus, predictions based on historic data may be a poor predictor. One way to account for this in the regression is to test whether there is a systematic trend over time. If so, future values of this trend variable can be inserted to reflect continuation of this trend. (Of course the hazard here is that past trends may not continue at the same rate into the future.) Another way to account for these changes is to incorporate data, such as on population health, that are likely to be the true underlying factors for amount of travel.

4.2 Assumptions of Regression Techniques

When we undertake a regression analysis, it is important to check whether the assumptions underlying the approach are at least approximately met. Essentially, this involves evaluating misspecification of the model, heteroscedasticity, multicollinearity, and true independence of the independent variables. In some cases, problems in regression violation are unlikely to affect your results much; in other cases, you may end up with incorrect estimates of the statistical relationships you want to uncover. Regression analysis makes the following assumptions:

First, the dependent variable is assumed to be a linear function of the independent variables, or it can be transformed (for instance, using logs) to a linear function. (Note, again, that the *coefficients* are linear, but the variables themselves may be nonlinear.) This was discussed previously.

Second, the distribution of sample observations around the predicted value of the dependent variable is the same for all values of an independent variable. This characteristic is called ***homoscedasticity***, and its violation, *heteroscedasticity*. Under heteroscedasticity, the variance around your prediction may increase as your independent variable either increases or decreases. Correcting for this problem often involves weighting the data or observations in relation to their variance. While it is unlikely to lead to major problems in your estimates of the coefficients, it could lead to problems in the estimates of your standard errors, and thus your t-statistics may be misleading.

Third, the values of each observation are independent of each other, and each independent variable is independent of the other independent variables. The individual observations are generally independent of one another in the case of survey data from hundreds of individuals, but they not be with time series data. The failure of this condition, known as *autocorrelation*, may (but does not always) lead to biased estimates of your coefficients. Additionally, the values of an independent variable are not always independent of the values of other variables. For instance, age may be correlated with wealth, or wealth may be correlated with educational attainment. Violation of this assumption is *multicollinearity*, discussed above under Regression Step 6.

Fourth, only the dependent variable is regarded as a random variable, and the independent variables are independent of the dependent variable. The values of the independent variables are assumed to be known with certainty. For example, if regression analysis is used to estimate the demand for a good when the price is $2, the true quantity purchased at this price can be predicted subject to confidence limits, but the price ($2) is known precisely. Additionally, that price should not be affected by the quantity demanded. For an individual consumer, the price is likely to be exogenous – you rarely have control over the prices you are charged at a store, for instance. However, if you are a major purchaser of a good and can negotiate over both the price and quantity of a good that you buy, then price is not an independent variable: it is indeed another dependent variable. In this situation, a linear regression with quantity as the dependent variable and price as the independent variable may give you very poor results.

This chapter cannot possibly cover completely all the issues you need to understand for a regression. Our major hope in reviewing these assumptions is to flag for you places where you may need to seek advanced help, either from a statistical consultant or from a statistics or econometrics text.

5. AN EXAMPLE: TESTING FOR ENVIRONMENTAL RACISM

Following the example we presented in the modeling chapter, we present the multiple regression results of the environmental equity model.

In this model, the hypothesis is that the race or ethnicity of the population of an area at the time of the permitting had a positive and statistically significant influence on the number of stationary air pollution sites permitted. The dependent variable is the density of stationary sources of air pollution permitted in the Denver, Colorado region, and race and ethnicity at the time of the siting decision are the two critical independent variables. However, the theoretical model suggests that a number of other factors might also influence a developer's decision about selecting this particular area for the air pollution source. These other variables could include such factors as availability of low cost land, the level of local opposition, and the percentage of total housing units that are rentals (since a high proportion of rentals – indicating that landlords live elsewhere -- would reduce homeowner opposition to siting of pollution sources). Including these variables in the regression improves both the goodness of fit of the equation and the ability to identify the effects of race and ethnicity separately from other factors.

Table 6-2 presents the results from a multiple regression with number of newly permitted stationary sources of air pollution per square mile of the zip code area as the dependent variable. The independent variables can be put in the following groups: percentage of the population that is Asian, Black, Hispanic and Native American (measuring the race/ethnicity effects); percentage with college degree and median household income (measuring other factors that may influence the nature of the neighborhood), median home value (measuring land cost), and percentage of housing units that are rentals (measuring homeowner involvement in an area).

As can be seen from the t-statistics in the third row of the table, only income and rental are statistically significant predictors at the 5% level (T-statistics > 1.96) of number of newly permitted stationary sources of air pollution in a given zip code. Thus, it appears that developers are less likely to propose, and cities less likely to approve, point sources in high income neighborhoods (as indicated by the negative coefficient), while these new sources are more likely (as indicated by the positive coefficient) to appear in neighborhoods with a high percentage of the population in rental housing units. Once we account for these two variables, ethnicity is not statistically significant. That is, low income, high rental white neighborhoods would be just as likely to be permitted for stationary sources of air pollution as would be neighborhoods of color. (Though not quite significant at the 5% level, the percentage of Blacks in an area seems to *decrease* the density of new sites, while more education, also of borderline statistical significance, seems to *increase* the density of new sites.) Overall the regression equation does explain 51% of the variation in the number of point sources in each zip code.

Table 6-2. Regression Results on Environmental Racism

	Coefficient	t-Statistic
Constant	-0.275	-0.263
Asian	0.029	0.263
Black	-0.02	-1.85
Hispanic	-0.011	-0.759
Native American	-0.002	-0.008
College	0.023	1.681
Home Value	5.52E-06	0.797
Income	-4.62E-05	-2.369
% Rental	0.019	2.026
R^2	0.51	

Thus, this study does not provide evidence that the race or ethnicity of an area leads to increased siting of air pollution sources in Denver, though it does suggest that low-income neighborhoods are more likely to be blighted with these facilities. Further analysis of the results, such as possibly combining the ethnic variables into one "minority" category (to examine possible effects of multicollinearity if minority groups tend to cluster together), may lead to some additional understanding of these results.

6. SUMMARY

This chapter has provided a very brief overview of linear regression, a very powerful tool for both testing of hypotheses (as in the environmental justice example above) and forecasting results (as in some of the discussions of recreation demand models, above). Our purpose here is simply to expose you to the usefulness of this method for analyzing data. Anyone who uses regression techniques should appreciate the importance of using this method thoughtfully. Such thoughtful use includes thinking about what data should (or should not!) be included, what variables are independent and which are dependent, choice of functional form, and careful interpretation of the results and predictions. The skills from this chapter will, we hope, enable you to understand, use, and critique others' use of regression analysis as well as to run your own regressions. At the same time, this chapter is only a brief overview. If you suspect your regression might violate some of the assumptions for a linear regression, or if you want more insight into the subjects discussed here, please consult additional sources.

Chapter 7

Principles of Benefit-Cost Analysis

One of the most widely used (and sometimes misused) forms of policy analysis is benefit-cost analysis (BCA). Most policy analysts end up spending time with BCAs, either contributing to them, creating them, or reading those conducted by others. BCAs frequently receive a great deal of attention in environmental policy debates, as they attempt to measure the net gains of a policy to society. Understanding what a BCA is, how it is conducted, what factors go into one, what factors should *not* go into one, and what are its limitations, are critical skills for any policy analyst.

This chapter describes the basic principles of benefit-cost analysis: the reasons for using BCA, what are considered benefits and costs, and limitations associated with BCA. Chapter 8, which discusses construction of discounted measures such as net present value, and Chapter 9, on nonmarket valuation of environmental quality, should be read following this chapter to provide a complete picture of what is involved in a complete benefit cost analysis.

1. WHAT IS BENEFIT-COST ANALYSIS?

Benefit-Cost Analysis (BCA) is a process of comparing, in common units (typically dollars), all of the gains and losses resulting from some action. The benefits are added together, and the costs are calculated as well. If the costs are subtracted from the benefits, the result is a net present value (NPV); if the benefits are divided by the costs, the result is a benefit-cost ratio (BCR). A BCA therefore summarizes the positive and negative aspects of an alternative into one number. A complete BCA compares alternative actions, through comparison of those index numbers, to determine which one provides society with the greatest net benefits through the most economically efficient use of its resources.

A BCA should include *all* the benefits and costs associated with an action, whether those goods are marketed (and therefore have price tags), or whether they are outside normal market operations (for instance, endangered species, air quality, or climate change). The BCA should also include both current and future values of all benefits and costs. How BCA weighs present and future values is discussed in Chapter 8 on discounting. How BCA can incorporate nonmarketed environmental resources is discussed in Chapter 9.

BCA can be viewed as information on the economic efficiency of a project or policy. This information is combined by public officials with information on equity, political feasibility, and other factors to determine the alternative that best meets society's needs. In some cases, BCA can be (or has been) used as a decision *rule* as well: whichever alternative provides the greatest net benefits to society is the preferred alternative. As will be discussed further, this second use of BCA is far more controversial than the first.

Regardless of exactly how the BCA information is used, the information can allow the public, agency directors, and elected public officials to separate out truly inferior alternatives from those that merit serious discussion. A BCA done following the techniques presented in this chapter will shed much light on four central features of an environmental policy or project. These are: (1) the efficient level of protection or environmental quality that balances the benefits and costs of additional protection or clean-up; (2) the optimal mix of environmental features in the alternative (e.g., should the light rail program have feeder transit buses or not); (3) the optimal size or scale of the project or policy (e.g., just how many miles of light rail line is economically justified); (4) the optimal timing of a when to implement the components of the management action or policy (e.g., when should each phase of the project or policy be implemented or built to maximize net benefits).

Even when BCA is used as a decision rule, it is not limited to simply determining the acceptability or unacceptability of alternative resource actions. In many cases a thoughtfully done BCA can help to isolate the economically inefficient features of a policy so that they can be dropped, and the economically efficient features can be implemented. Thus BCA can be used as a guide to designing the most beneficial projects or policies by subjecting each separable feature to BCA. Along the same lines, BCA can provide information on approximately when a project or resource management action should be undertaken to maximize benefits, and it can provide information on the distributional effects of an action as well.

2. USE OF THE "WITH AND WITHOUT" PRINCIPLE

As we mentioned in Chapter 2, it is important to scope out the problem, and one critical element of that is establishing what the baseline, or the effects without the policy, will be. The effects of the policy are then compared to this future state of the world without the policy to establish the net effect of the policy. This avoids using a "before vs. after" viewpoint, which may attribute some changes that occur at the same time as the policy implementation to the policy, when in fact the changes were already underway. That is, just because nothing is done does not the mean the future without-project level equals the current level. The "with-without" principle is also very helpful in zeroing in on exactly what resources need to be fully evaluated in a BCA, and which can be ignored. If the level of a particular environmental effect would be the same in the with and without alternatives, then it need not be analyzed as part of the BCA. The bottom line in a BCA is driven only by changes in the quantities (and qualities) of those effects.

3. CONCEPTUAL FOUNDATION OF BENEFIT-COST ANALYSIS

One overall objective of a BCA is to determine which combination of resource uses produces the greatest net gain to society. These gains and losses are measured from the viewpoint of individual members in society. Since people prefer more benefits to less, reallocations of resources that increase the benefits to one individual without reducing the benefits to anyone else would be preferred. The situation where nobody loses from a policy, and at least some (and perhaps all) gain, is known as a Pareto improvement. It is difficult (though not impossible) to object to a Pareto improvement.

In many cases, though, even if a policy has net benefits, some gain at the expense of others. For instance, building a reservoir provides benefits to those who like lake-based recreation, but it harms those who prefer river-based recreational activities. In this case, if those who enjoy lake-based recreation can adequately compensate those who like river-based recreation – for instance, by paying for the extra expenses the river people would incur to get to other rivers – then both parties at least come out even, if not ahead, from the development of the lake. If it is possible for the gainers to compensate the losers fully, the project is *potentially* Pareto improving. For a Potential Pareto improvement, compensation need not actually happen, but the gains are large enough that it could happen. Those who end up losing in this situation have understandable reasons to object to that alternative, since they bear the costs without reaping the benefits.

3.1 Defining what are Benefits

To actually determine if a policy change represents an improvement in social well being, you must be able to actually measure the net gain in benefits. Much like beauty, the benefits of a policy are often in the eye of the beholder. Are the jobs created by the project a benefit? Are the additional commodities produced an economic benefit? What about cost savings to society from policies that encourage telecommuting instead of automobile commuting? The answers to these questions depend on the analyst's valuation philosophy and accounting stance.

3.2 Valuation Philosophy

The anthropocentric view on which BCA is based states that people are the ultimate source of value. Goods and services provide benefits only if individual people are made better off. The magnitude of the benefits received is determined by each individual's own judgment of how much better off s/he is. One conceptualization of "how much better off" is to look at the change in what economists call the individual's level of utility (e.g., level of satisfaction). Unfortunately, an individual's level of utility is not directly observable, although we can indirectly determine the utility provided by observing the actual choices made by individuals in the face of different prices or incomes.

3.3 Money as a Metric in BCA

A BCA requires the ability to add up different kinds of benefits and subtract off different kinds of costs. For instance, a benefit of reducing air pollution might be increased visibility and improved human health, while the cost might be an increase in the price of electricity due to higher pollution control expenses at the electrical utility. To assess the overall net benefits of reducing air pollution, analysts must be able to add up visibility and human health and subtract off pollution control costs in a way that makes sense. One way – the method underlying BCA – is to determine the value per unit associated with each benefit or cost, and multiply each unit of the good by that value. For instance, if improving visibility from 1 to 5 miles has an average value to each affected person of $10 per year, and there are 1 million affected people, then the visibility benefit would be $10 million. When a similar calculation is done for all benefits and costs, the benefits can be added up, and the cost subtracted, to find the net benefit associated with improved air quality.

Monetizing environmental quality, such as air and water quality, is often considered controversial, for reasons ranging from the fact that we're unfamiliar with thinking of these goods in market terms, to the objection that placing price tags on these goods demeans them in the same way that assigning price tags to human emotions is considered absurd. At the same time, *not* including them in a BCA will lead to a distorted answer in the analysis. Additionally, while people may be reluctant (for a range of reasons) to price environmental goods, it is probably fair to say that, for many people and for most environmental goods, there is a finite amount they would be willing to pay for very high levels of environmental protection. When economists seek to estimate these values (as will be discussed further below), the intention is to summarize, to the best of their abilities, the tradeoffs that people face in protecting the environment. The primary role of the dollar metric is to provide a means of comparing apples and oranges (or visibility and pollution control technology) in a way that enhances decision-makers' understanding of the tradeoffs.

3.3.1 From Utility to Income

By graphing a consumer's utility using what are called indifference curves, it can be shown that a change in income can be used to measure a change in utility. Thus one measure of how much an individual benefits from an improvement in air quality is the maximum amount of income s/he would be willing and able to pay for the improvement rather than go without. This is the sum of money where the individual is indifferent between paying that sum of money and having a given increment in air quality (e.g., 5 more miles of visibility) versus keeping the money and foregoing the gain.

3.3.2 Income as a Proxy for Sacrifice of Other Goods

We will illustrate this concept of willingness to pay more concretely later in this chapter. At this point we wish to stress that willingness to pay (WTP) in the form of income is simply a proxy for willingness to give up other goods and services to have the policy or project under study, and it therefore serves as a measure of the benefit of a good. In essence, maximum willingness to pay is often driven as much by a willingness to substitute the new good for the old good as it is by level of income. For example, a person that really likes whale watching might be willing to give up three trips to the open-air concert at the downtown park to have one additional whale watching trip. In this sense, income influences willingness to pay only because we use dollars as the measuring units, not because the individual is directly foregoing income. Rather, the individual is foregoing three concerts to gain one whale watching trip. But if we wish to have a dollar measure of the three open-air concerts foregone to gain one whale watching trip, we must convert the concerts in the park foregone into their income equivalent in dollars. In this way we can compare the value of the whale watching trip gained to the cost society must incur to provide the whale watching opportunities (perhaps reduced off-shore oil drilling or commercial navigation).

3.3.3 From Individual Benefits to Social Benefits

Now that we have an initial means of measuring benefits from an individual's point of view, we need to see how this information would be used to determine if some policy or management action is desirable. To do this, we assume that what is best for society is what is best for the aggregation of individuals in society. That is, society's benefits are the simple sum of the benefits individuals derive. For example, assume society is made up of just two people, Mr. Loomis and Ms. Helfand. There are two possible bundles of goods that cost the same to produce. Society is attempting to decide whether the highest valued use of resources is to produce Alternative B rather than the current Alternative A. Table 7-1 shows the benefits to each of them with Alternative A and Alternative B.

Review of Table 7-1 shows that Alternative B is preferred since Ms. Helfand receives a bundle of goods she values more highly (i.e, is willing to pay more for) than in Alternative A. At the same time, Mr. Loomis is neither better off nor worse off. Thus, moving from Alternative A to B results in an actual Pareto Improvement in society's welfare: Ms. Helfand gains ($25) and Mr. Loomis stays the same. There is an actual clear gainer and no losers. Thus the 'economic pie' gets larger, with Ms. Helfand's slice growing and no change in Mr. Loomis's slice.

Table 7-1. Consumption Benefits with Different Bundles of Goods and Services

Alternative	Ms. Helfand	Mr. Loomis
A	$100	$150
B	$125	$150
C	$40	$200

Once we have adopted alternative B, we have obtained a Pareto Optimum, or optimum economic efficiency. That is, there are no other alternative allocations of resources that make one of them better off without making the other worse off. Clearly alternative C cannot do it, as Mr. Loomis gains $50 but at the expense of Ms. Helfand losing $85 as compared to Alternative B. At alternative B the economic pie is at its maximum size, given the current technology and tastes and preferences of Mr. Loomis and Ms. Helfand.

3.3.4 Potential Pareto Improvements

Unfortunately, many policies or management actions, such as raising a gasoline tax or permitting a subdivision in what is currently open space, will result in some groups in society losing and others gaining. For example, is Alternative D in Table 7-2 better than Alternative B? Alternative D does not represent an actual Pareto Improvement as Mr. Loomis (current homeowner adjacent to the empty field) is made $25 worse off (compared to B), while Ms. Helfand (the developer) is made $75 better off (compared to B). What can we say about an alternative policy that makes one person better off but another worse off? If we can add individual valuations of Ms. Helfand and Mr. Loomis together we can determine if the sum or total of the valuations are the highest with Alternative D. Using the criterion of total net benefits, Alternative D is preferred since it yields the greatest value of total output. In other words, the economic pie is larger than with any other alternative.

Table 7-2. Identifying a Potential Pareto Improvement Using the Total Benefits Rule

Alternative	Ms. Helfand	Mr. Loomis	Total Benefits
A	$100	$150	$250
B	$125	$150	$275
C	$40	$200	$240
D	$200	$125	$325

Alternative D represents a Potential Pareto Improvement: the gainer (Ms. Helfand) could potentially compensate the loser (Mr. Loomis) and still have a gain remaining. With Alternative D in Table 7-2, Ms. Helfand could afford to give Mr. Loomis (or the City Parks Department to preserve some other nearby open-space) $25 to get him back to his original level of benefits ($150 in Alternative B), and Ms. Helfand would still be better off by $50. On this basis, Alternative D is the most economically efficient alternative. In no other alternative could the beneficiaries compensate the losers and still have any gain left over. This is sometimes referred to as a compensation test. Such tests can be quite elaborate, including not only whether the gainers can fully compensate the losers and still have a gain left over, but whether the losers are unable to bribe the gainers into foregoing the program. (See Mishan, 1981).

For the total or sum of benefits to reflect a meaningful measure of social well being involves at least two key assumptions: (1) that we can meaningfully compare different people's valuation of goods or services; (2) that a dollar's worth of benefits provides the same change in individual utility or well being to every person.

3.4 Economic Efficiency and Equity in Computing Total Social Benefits

These assumptions may be a bit troubling. There are certainly cases where a hamburger to a homeless person who gets one meal a day provides more satisfaction to that person than the same hamburger provides to a rich person who eats three meals a day. But our measure of value is the amount of sacrifice a person would make to obtain the good in question. Because the poor person has less other goods (or income) to sacrifice, his or her revealed value in the form of willingness to pay would likely be lower. As cruel as this extreme example sounds, we must remember that BCA is about economic efficiency, not equity. This is the logic of the marketplace. All dollars have equal power in the market. (This example also highlights the importance of not relying solely on BCA to make a decision, as equity clearly matters to many people.)

As in evaluating market outcomes, society certainly is concerned about equity when ranking alternatives in BCA. As Chapter 3 explains, how the total benefits are distributed may be as important to social well being as the amount of total benefits. The distributional implications of economically efficient resource allocations are best handled by displaying not only the total net benefits, but also how the benefits are distributed across different income, age, geographic, or ethnic groups, or other groupings. In this way the political system can choose to weight benefits to different groups differently if it so desires. Since any differential weighting system is a value judgement, it should be reserved for the political system and not be carried out by the analyst unless the analyst is explicitly given weights for this project by the client.

In the absence of explicit, project-specific weights, we will follow standard practice used in federal Benefit-Cost procedures (U.S. Water Resources Council, 1979, 1983) in this book and weight all individuals' dollars equally. While this is just one of the many possible weighting systems, it is certainly one consistent with much of the environmental policy legislation. Recent Executive Orders on Environmental Equity, in fact, stress *equal* treatment of all groups in society, not preferential treatment.

3.5 Would We Always Want To Pay Compensation?

Another concern often expressed about reliance on Potential Pareto Improvements rather than an Actual Pareto Improvements is the failure to actually pay the compensation to the losers. There are several reasons for not requiring actual payment of compensation even if failure to pay compensation actually leaves some people worse off. These reasons may be grouped into ones relating to economic efficiency and equity. The economic efficiency reasons are based in the idea that much of the net gain in benefits to society might be consumed in attempting to carry out the steps necessary to actually pay the compensation. These include:

1. The high information costs associated with identifying the specific individuals that gain and the specific individuals that lose.
2. The large transactions costs of facilitating actual payments from the gainers to the losers.

The equity reasons for not paying the compensation include:

1. Payment of compensation could worsen equity if gainers are poor and losers are relatively well off. Certainly, having the poor compensate the rich would run counter to most accepted norms of equity.
2. A progressive tax system (i.e., one in which tax rates rise with the level of income) coupled with an effective system of transfer payments (e.g., welfare, food stamps, rent subsidies, etc.) will result in partial compensation of the losers as gainers must pay an increasing portion of their gain as taxes and losers' taxes will be reduced (Sassone and Schaffer, 1978:11).
3. If BCA is applied to most government programs, everyone will eventually be a gainer in some evaluations. We simply recognize that, while our piece of the expanding economic pie might get smaller from one project, if the economic pie is continually expanding as resources are moved from low valued uses to higher valued uses, eventually our slice will increase (Sassone and Schaffer, 1978: 11).

Nonetheless, it is becoming more common for communities to ask for compensation from developers when the developer wishes to impose a locally undesirable land use, or LULU, on a single community when the LULU benefits society as a whole. If there are really net benefits to society from utilizing that location, there should be sufficient resources to pay compensation. Whether to pay the compensation affects the distribution of the benefits and costs but does not affect whether the project has net economic benefits. Be warned that developers will often say that paying the compensation would make the project uneconomic. If this is really true, perhaps the project has minimal economic benefits to start with, and society would be better without the project; after all, if there are net gains to society from the project, those who gain from the project should be able to compensate those who lose. Making a developer pay the compensation can sometimes provide a good test for just how large are the net economic benefits of that particular location.

4. ROLE OF ACCOUNTING STANCE IN DETERMINING TRANSFERS

While benefits are defined from the viewpoint of individuals, often one person's gain is exactly offset by another person's loss. That is, some resource reallocations do not represent net gains in economic efficiency, but are rather simply a transfer of economic activity from one person to another or one location to another. For example, if Ms. Helfand receives $100 worth of cheaper electricity from a new hydropower reservoir but Mr. Loomis loses $100 of river recreation downstream when the reservoir is in operation, then there is no net gain to this society.

While this seems very straightforward, if we add one element of realism to this example, we can illustrate a frequent confusion over what constitutes benefits. Assume that Ms. Helfand (who gets the cheaper electricity) lives upstream in the State of Kentucky, but Mr. Loomis lives downstream in Tennessee. If you as the analyst measuring the benefits of constructing the new hydropower reservoir worked for the State of Kentucky and took a state view, then you would find the reservoir resulted in a gain of $100 of benefits to residents living in Kentucky. This state viewpoint is one accounting stance. By accounting stance we mean the geographic viewpoint of what benefits and costs count or matter, and which ones do not matter. Thus with the State of Kentucky our accounting stance, only benefits received in Kentucky and only costs incurred by Kentucky residents would count. The state analyst would ignore both benefits and costs occurring outside our region of interest.

From a national accounting stance such a state accounting stance is too narrow to reflect all of the benefits and costs to all people. That is, the lost river recreation occurring in Tennessee is a real decrease in well being and should be counted. A state accounting stance in this two-state example would result in failure to incorporate a negative externality of the reservoir into benefit-cost calculations. Since one of the reasons for public regulation of resources is to internalize such externalities into public decision making, clearly at least a national accounting stance is required to insure that the complete benefits and costs of a resource management action are reflected. As discussed in Chapter 1 on laws relating to environmental policy analysis, for hydropower relicensing and regulatory impact analysis, usually a national accounting stance is the appropriate one for most actions. This is especially true when evaluating *national* policies for control of pollutants that move across state lines via winds or rivers, such as the Mississippi, Missouri, Colorado, or Columbia Rivers. It is also true of regulations like airline safety, which affect people living across the nation. It is also appropriate to use a national accounting stance when dealing with state programs that use federal cost-sharing funds. These federal funds reflect taxes paid by households across the entire country.

But there are times when even a national accounting stance is too narrow. Consider the case of controlling acid rain in northeastern United States and southeastern Canada. The costs of controlling acid rain are incurred largely in Midwestern States. The benefits are received downwind in New England and in Canada. The benefits from acid rain control the Canadians would receive are real increases in their social well being. They should count when we add up the benefits to compare them to the pollution control costs. Once again, if analysts take too narrow an accounting stance, real benefits to individuals (who just happen to reside in a different political jurisdiction) will be ignored. The U.S. might erroneously reject acid rain controls because the benefits (as measured as the gain to just U.S. citizens) are less than the costs. But the true social benefits (including the Canadians) might well exceed the costs.

The general guidance to the analyst is to adopt the spirit of the first legislation requiring BCA and measure benefits and costs "to whomsoever they may accrue". Much like dealing with equity in BCA, concerns about which states or nations gain and which ones lose are best dealt with by displaying the distribution of benefits and costs to each political jurisdiction. This is much better than adopting a narrow accounting stance which results in complete omission of certain state's or nation's benefits and costs. Such an omission is often a recipe for economic and ecological disaster.

5. GROSS WILLINGNESS TO PAY, COST, AND NET BENEFITS

It is now time to begin to answer one of the first questions posed by this chapter: is the additional employment created by a resource management action a benefit or a cost? Often times one reads in the newspaper that a new power plant, factory, or relaxation of environmental regulation will "create" a certain number of jobs. The same newspaper article might quote a timber industry official who states that preserving an area as endangered species habitat results in society foregoing trees that are worth one million dollars at the lumber yard. Is this an accurate statement of the opportunity cost of designating the area as endangered species habitat? We now turn to an example that will allow us to answer these questions.

Table 7-3 presents data on two resource uses that could take place on a specific parcel of land that is currently a wetland. One of the current uses of the wetland is as a habitat for waterfowl, which allows for a "duck club" or waterfowl hunting. Of course the wetland provides other non-market functions such as groundwater recharge, storage of storm waters from city streets, and detoxification of heavy metals from the stormwater. We will return to these non-market benefits at the end of this example. The proposed use is to drain the wetland and have livestock grazing. For this example, we assume the owner has petitioned you, as an analyst with the county, for a permit to drain the wetland for livestock grazing. The column headings relate to Gross Benefits, Cost of Production (or Expenditures), and Net Benefits. Net Benefits is equal to Gross Benefits minus the Cost of Production.

Table 7-3. Example of Annual Gross and Net Benefits

Resource Use	Gross Benefit (Price * Quantity)	–Cost or Expenditures	= Net Benefits
Livestock Grazing	$95,000	$45,000	$50,000
Waterfowl Hunting	$60,000	$20,000	$40,000

We can use Table 7-3 to answer the question about the economic effects to society from maintaining the wetland area for waterfowl hunting as compared to draining it for livestock grazing. If we keep the area as a wetland and have a waterfowl hunting operation, what does society forego? Suppose the rancher would receive $1 per pound for the 95,000 pounds of beef raised on the drained wetland. But this is not the net value of the beef to the rancher or to society if we allow this operation. The rancher must spend $45,000 purchasing his initial cattle stock, feed supplements, veterinarian services, and labor to tend the cows and truck them to market. The net gain to the rancher and the nation is only the difference between what society gets ($95,000) and what it gives up ($45,000). The net benefits of $50,000 represent the gain or net income to the rancher over and above the costs. Suppose that $30,000 of the $50,000 in costs is wages paid to hire 2 workers to move the livestock from where the young calves were purchased to the pasture, move them from pasture to pasture, replenish salt licks, repair fences and truck the full grown cows to market.

Now let's examine maintaining the wetland for waterfowl hunting. For the moment, assume 200 hunters would pay $300 each year to go waterfowl hunting in this area. That is, they would pay $300 each year to have the area available for waterfowl hunting rather than go waterfowl hunting at the next best area. This yields a gross benefit of $60,000 in Table 7-3. However, the landowners incur costs of $20,000 for hiring one worker ($15,000) and materials ($5,000) each year to erect the hunting blinds at the beginning of each hunting season and then to take them down and store them each year after the season is over. Thus, the landowner receives a net income of $40,000 a year from the waterfowl hunting operation, which is decidedly less than the livestock grazing operation.

So what is the cost to landowner and society from denying the permit to drain the wetland? It is not the $95,000, worth of beef, as the landowner and society saves $45,000 in costs from not producing the beef here. Even the net income of $50,000 is not lost, as $40,000 a year can be made in the waterfowl hunting operation and still protect the wetland and its waterfowl. Thus the net cost to the landowner is $10,000, the difference between the $50,000 with livestock grazing and $40,000 with the waterfowl hunting. Correct comparison of net benefits in the "with permit" versus "without permit" scenarios identifies the opportunity cost of protecting the wetland and its other non-market ecosystem services, such as groundwater recharge and stormwater retention, as $10,000, not $95,000 worth of beef or even $50,000 of lost income, as there was another economically viable activity (although certainly not as profitable to the landowner). Thus protection of the wetland and the other ecosystem services provided by the wetland would have to be worth at least $10,000 each year to make this an economically efficient decision to society.

But what about the differences in wages paid and employment, you might ask? If the wetland was drained for livestock grazing, two workers would have been employed, while with waterfowl hunting only one is. But, if there is a market demand for the beef that would have been raised on the drained wetland, and we deny the permit, another ranch will simply expand its production by the additional cows to meet that demand. If there really is a market demand for that beef, the private market will provide an incentive to ranchers to produce it. Where these additional cows are raised, the same two workers will be needed to pick up the calves, tend the cattle, and truck them to market. Thus two more workers will be needed somewhere else, possibly in the same region. And this makes sense, since those two ranch hands that would have worked for the landowner who wishes to drain the wetland are unlikely to be unemployed for ever! If they like ranching, they will move to where the two new jobs are. If they like that specific area, one might work for the landowner at the waterfowl hunting operation, and the other would look for work elsewhere.

While some State Fish and Game agencies would like to count the amount hunters themselves spend as a benefit of hunting, it is clearly a cost. Certainly, the expenditures by hunters on ammunition and transportation would not be lost to the economy as a whole if the wetland were drained, any more than the expenditures on feed supplements would be lost if the wetland is not drained. The ranchers expanding production of their herds will buy the same amount of feed supplements. The only way the hunter or rancher expenditures could be lost to the economy as a whole is if they went home and set fire to the money! Since people are unlikely to do this, what actually occurs is that hunters spend that money on their next most preferred activity or area if the wetland is drained.

To see how decisions not to permit draining a wetland or building a subdivision or landfill in one particular location is just a transfer of economic activity, consider the decision on location of the Intermountain Power Project (IPP) coal fired power plant. In 1978 this large coal fired power plant was proposed at Hanksville, Utah, near Capitol Reef, Canyonlands, and Arches National Parks. Due to concern over air pollution reaching these National Parks, it was decided to build the power plant on the west side of Utah, near the town of Lyndal and far away from any National Parks. While the town of Hanksville, Utah lost all of the influx of construction workers and plant operations personnel, this was perfectly offset by the town of Lyndal's gain of the same level of construction workers and plant operations personnel. Viewed from an international, national, or state accounting stance, there was no change in employment of resources. But viewed from a narrow accounting stance of the city of Hanksville, there was a net loss. Viewed from the narrow accounting stance of the city of Lyndal, there was a net gain for them. As can be seen,both of these narrow accounting stances seriously misrepresent the change in economic social well being. The same number of workers were employed to build and operate the power plant regardless of its location. Thus new projects in one location do not, at the national level, "create jobs" that would not have otherwise been demanded. Market demand creates the jobs.

5.1 Assumptions about Employment of Resources

As discussed above, employment of labor, materials, diesel fuel, machines, and other factors of production represent opportunity costs to society. That is, use of a given amount of our scarce resources in one project or management action necessarily means giving up using these resources to produce output from some other project. All of this assumes reasonably full employment of resources. If, in fact, the labor employed in one project would have been otherwise unemployed for the duration of its employment in the project under study, then there is no opportunity cost of using that labor in this project. That is, use of unemployed resources does not require society to give up anything else to gain the new level of output.

Just how likely is it that the resources employed by a project would otherwise be unemployed? If the project is a long lived one, requiring labor and materials over several years (or even decades), it is very unlikely that one could rely on unemployed resources. Excess unemployment (above the background frictional levels) tends to be transitory or existing for short periods of time during recessions. Most resources, whether labor or machinery, are relatively mobile and will seek out employment rather than remain unemployed for years at a time. However, the possibility exists that, during some phase of a public project, otherwise unemployed resources would be put to work. If this is the case, the analyst can either count the economic costs of such resources at zero (or near zero, depending on the value of unemployed leisure time to workers) or, alternatively, if labor costs (measured as the opportunity cost of that labor) are included on the cost side, also count the wages paid as a project benefit. Either treatment will reflect the appropriate change in a project's net benefits.

5.2 Displaying Changes in Employment and Economic Activity

Often laws such as the National Environmental Policy Act, which governs preparation of Environmental Impact Statements, or local politicians require that the employment related to a public project or policy action be displayed. Thus, it may be appropriate to translate direct changes in project employment into total employment by including spin-off employment in support sectors of the economy induced by the project. In the same way it may be important to calculate the change in personal incomes in a given county or region that results from an increase in economic activity induced by the project. While both the induced employment and personal income generated by a project that uses fully employed resources are a transfer, there is often interest in knowing the magnitude of those transfers.

This economic impact analysis often serves at least two useful purposes. First and foremost, it can help local government planners plan the infrastructure of roads, schools, hospitals, housing, and parks that will be needed to accommodate the additional workers and their families. Second, politicians may desire to stimulate selected regional economies at the expense of other areas. For example, expanding job opportunities in rural areas and increasing the tax base in rural economies may be a distributional goal. Even though this rural gain may be offset by a loss of the same economic activity in an urban area, politicians may wish to stimulate rural economies to maintain their viability. The techniques for computing income and employment effects of projects via multiplier analysis (using input-output models) are discussed in Chapter 10.

6. MEASUREMENT OF ECONOMIC EFFICIENCY BENEFITS

Now that we have identified what we wish to measure, the next step is to determine how to measure it. The methodology chosen to measure the benefits of resources must allow us to compare marketed resources and nonmarketed resources. To obtain consistency in valuation for both marketed and nonmarketed resources, economists rely on values measured from consumer's demand curves and businesses' supply or cost curves.

6.1 Benefits to the Consumer

It is easiest to illustrate valuation of a resource with an example. Consider Figure 7-1, which presents a consumer's demand curve for gasoline. At a price of $2 per per gallon of gasoline, the consumer wishes to purchase six gallons of gasoline each week. What are the benefits of being able to buy six gallons of gasoline at $2? One's first reaction might be that, if the consumer paid $2 for each gallon, they must be worth $2. But if we read up from the quantity axis, to the demand curve and over to the vertical price axis, we see that only the sixth gallon purchased is worth just $2. The fifth gallon is worth $3, the fourth is worth $4, the third one $5, and so forth, to the first gallon each week being worth $7. That first gallon of gas provides transportation for important trips and errands that cannot be accomplished by substitute transport such as bike or bus. Thus it provides $7 worth of benefits. That is, if the consumer could only have

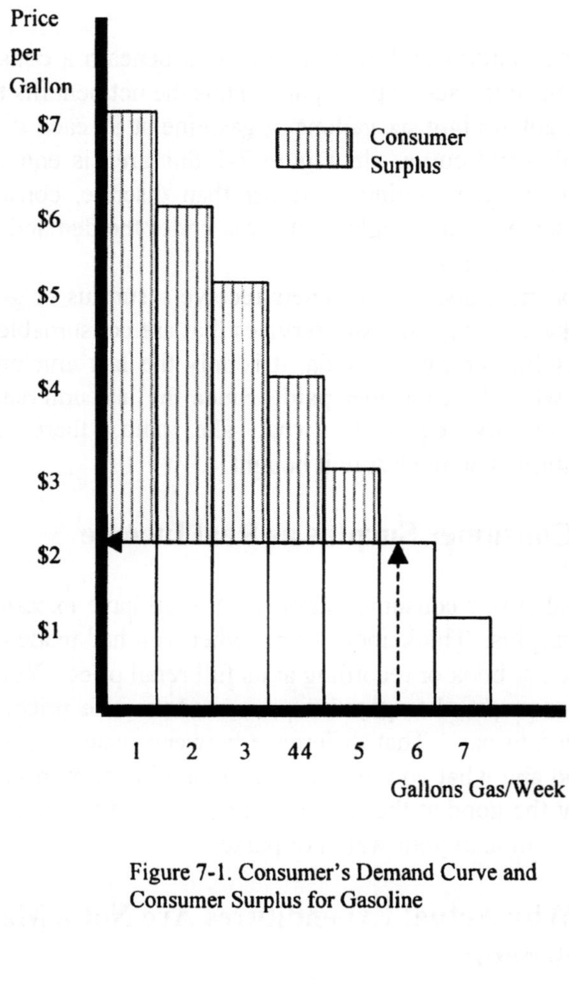

Figure 7-1. Consumer's Demand Curve and Consumer Surplus for Gasoline

one gallon of gas a week, she would pay up to $7 for it. Since the market clearing price is just $2, the consumer receives a "personal profit," or what economists call a consumer surplus, of about $5 on this first gallon of gasoline.

Consumer surplus is defined as the area beneath a consumer's demand curve but above the actual price paid. Thus the net benefits to the consumer from being able to buy six gallons of gasoline at $2 each is the striped area under the demand curve. In Figure 7-1 this area is equal to $15 (if the demand curve were continuous rather than discrete, consumer surplus is often represented as a triangle -- the area under the demand curve in excess of price actually paid).

Thus for most goods, consumers receive a surplus or gain in excess of what they pay. For goods and services that are consumable in small units, such as gasoline or cans of soda, it is only the last unit purchased that is worth just what the consumer paid. Since the last unit has a value to the consumer exactly equal to what she paid, there is usually no consumer surplus on the last unit bought.

6.2 Consumer Surplus as Real Income

How real is this consumer surplus? We all have experienced receiving consumer surplus. Think about a time when you had made up your mind to buy a particular book or recording at its full retail price. You went up to the cash register prepared to pay this price, and the sale price was lower than you expected to pay. That difference between what you would have paid for the good and what you did pay was a tangible gain in your real income. You bought the good at the sale price and retained the consumer surplus as added real income in your wallet or purse.

6.3 Why Actual Expenditures Are Not a Measure of Net Benefits

While use of consumer surplus as a measure of the economic efficiency benefits to the consumer seems quite straightforward, some people become confused over how to deal with the actual money spent by the consumer. A simple example will illustrate why the total amount of consumer expenditure on a good is often a poor indicator of the total benefits the person receives from consuming the good. At the price of $2 per gallon of gasoline the consumer buys six gallons, for a total expenditure of $12. However, if the price of gasoline rose to $6 per gallon, Figure 7-1 indicates the consumer would buy two gallons per week. This yields the same $12 total expenditure. Using the actual expenditure as a measure of benefits would lead to the conclusion that the consumer would receive identical benefits from consuming two gallons of gasoline as she would six gallons per week.

Common sense tells us the consumer would prefer to pay a total of $12 to get six gallons of gasoline per week rather than two. If we calculate the consumer surplus of these alternative combinations (six gallons of gasoline at $2 versus two gallons at $6) on Figure 7-1, we see that consumer surplus is a good indicator of the preferred combination. As shown above, the consumer surplus for six gallons of gasoline purchased at $2 is $15. The consumer surplus for two gallons purchased at $6 is only $1 ($7 minus $6 price). Thus, the consumer surplus makes it clear: six gallons of gasoline provides more benefits to the consumer than do two.

While this may seem like an incredible amount of effort to prove the obvious (and it is!), reliance on expenditures as a measure of benefits is, unfortunately, a common mistake in policy analyses. It is a particularly easy mistake to make when actual expenditures are readily observable in the marketplace and consumer surplus can only be inferred after statistically estimating the product's demand curve. The temptation is great to simply use expenditures, rather than go to the trouble of statistically estimating demand curves. Hopefully, the error of using consumer expenditures as a measure of benefits has been made clear.

Another pitfall in calculating consumer benefits can be illustrated by the question, "Is not the $2 per gallon or the $12 spent on all six gallons of gas part of the benefit of the gasoline?" The key here is to realize that, when oil becomes so scarce it is too valuable to make into gasoline, consumers would only lose their consumer surplus, not their expenditures. If people could no longer purchase gasoline at $2 per gallon, the $12 they are currently spending on gasoline would be spent on some other good. What would be lost to the consumer and society is the additional utility or benefits that spending the $12 on gasoline provided, that is not received from buying the next most preferred good (say, methanol fuel). Remember, demand curves are adjusted for the price and availability of substitutes, so that consumer surplus is the value over and above the next best substitute good. Thus, from the consumer's perspective, the $12 spent is a cost of gasoline, not a net benefit. However, the $12 is received by gasoline retailers and oil producers. Part of this $12 pays for costs of production, but part of it reflects benefits to the business producer. We now turn to measuring the benefits to the producer.

6.4 Benefits to the Producer

To begin with, we will simplify the analysis and measure benefits to a firm or business that is a price taker. A firm is a price taker when its output is a small part of a national or international market for that product. Continuing our gasoline example, one individual independent oil producer would be so small as to have no effect on the price of gasoline. Figure 7-2 illustrates the relationship between the market for gasoline (measured in millions of gallons) and the individual oil producer (whose output is measured simply in gallons of gas). The equilibrium price is set in a national or international market based on total consumer demand and total industry supply. The demand curve facing the individual producer is horizontal at the current market-clearing price for gasoline. The reason for this is the independent producer's contribution to supply makes up such a small fraction of the total supply of gasoline that doubling its production of gasoline or cutting production in half would have no effect on price. In addition, the independent oil producer can sell all it wants at the existing market price.

Besides the horizontal demand curve, Figure 7-2 shows the independent oil producer has an upward-sloping marginal cost curve. That is, as the producer attempts to pump more oil and refine more gasoline from a given size oil deposit in a given amount of time, its costs per unit of output rise as it must intensify management, add more capital inputs, and add more shifts of labor (and pay a premium for the difficult-to-fill night shift). The marginal cost curve is the independent producer's supply curve. It reflects the minimum dollar amount for which the producer would supply each additional gallon of gasoline.

To calculate the net economic efficiency benefits to the independent oil producer, we note that sales of the first 500 gallons of gas cost only $0.75 per gallon to produce, but the owner receives the market-clearing price of $1.50 per gallon. Thus the owner receives a producer surplus, or economic profit, of $0.75 per gallon on these first 500 gallons. The next 500 gallons of gas cost $1 per gallon to produce, but are again sold for $1.50 per gallon. The producer surplus on these units is $0.50 per gallon. It is only the 2000th gallon in which the price received by the producer equals the minimum supply cost, and therefore no producer surplus is received on the last unit produced. Thus, the total producer surplus for 2000 gallons of gas is $1000 ($1.50-0.50 times 2000 pounds with the product divided by two since producer surplus is a triangle). This is the shaded area in Figure 7-2. Notice the similarity in logic between derivation of producer and consumer surplus. In both cases, economic efficiency benefits are the net gain over and above the costs.

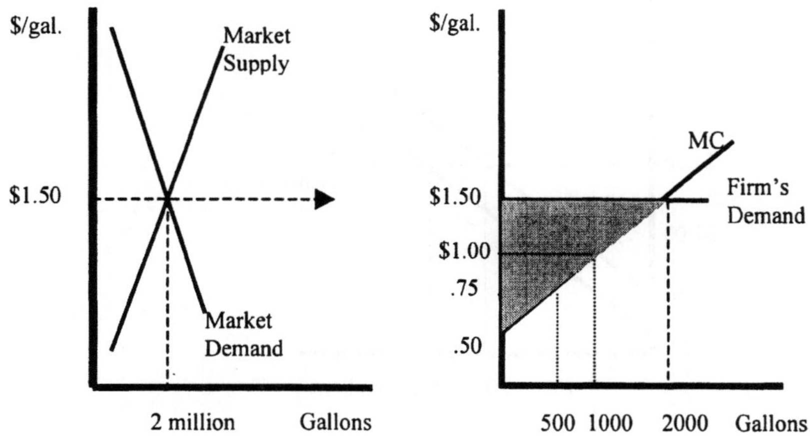

Figure 7-2 Market, Firm and Producer Surplus

Another means by which to calculate producer surplus and to see its relationship to a firm's profit is to recognize that the producer surplus triangle is the difference between a firm's total revenue and total variable cost. The total revenue to the producer is $3000 ($1.50 times 2000 gallons). The total variable cost is the area under the marginal cost curve up to the profit-maximizing level of output (here 2000 gallons). In this example the area under the marginal cost curve is $2000. Thus, the producer surplus is $1000 ($3000-2000), just as was calculated before. Note that, just as in the case of the consumer, producer surplus is the gain over and above the firm's actual expenditures. The firm's expenditures are the cost of inputs. The use of inputs by this producer results in an opportunity cost to society, since society must forego whatever else the inputs would have produced in their next best use.

Next let us examine the benefits of a relaxation of a county groundwater protection regulation that would reduce the cost of oil production to this, the only oil producer in this county. For example, instead of twice-daily drawing and analysis of groundwater quality samples to detect and any seepage of oil into the groundwater, monthly samples are being proposed. These effects would be translated into lower production costs and hence a downward shift in the producer's marginal cost curve.

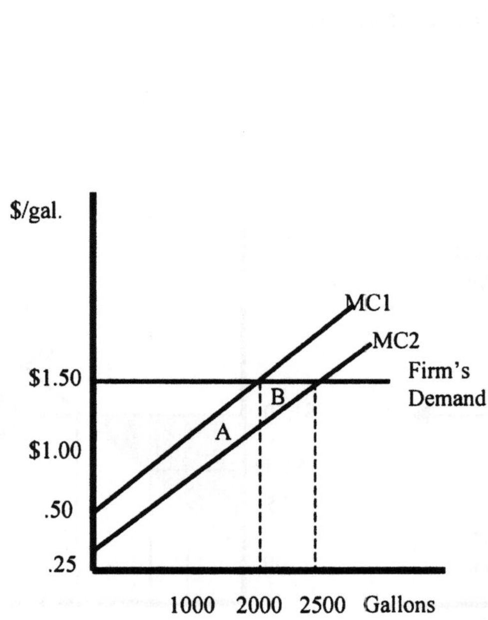

Figure 7-3 Change in Producer Surplus

As Figure 7-3 illustrates, this downward shift in the producer's marginal cost curve has two effects. First, the producer receives a resource cost savings, as the original level of output (2000 gallons of gas) can be produced at a lower cost. Thus, these resources can be freed up to produce other goods and services that society values. The resource cost savings on the original level of output is equal to $500 ($.25 times 2000 gallons). This is denoted as the trapezoidal area labeled A in Figure 7-3.

The second benefit to the producer and society arises because, with the lower cost of production, it is profitable for the independent producer to increase the rate and duration of pumping and expand refining to 2500 gallons. With the new marginal cost curve, the cost of producing the 2000th gallon is now less than the price. As such, a profit maximizing oil producer will expand production until the new marginal cost equals the market price. As such there is a gain in benefits equal to $62.50 ($0.25 times 500 additional gallons of gas divided by two since producer surplus is a triangle). This is the triangular area labeled B in Figure 7-3. Another way to view this second gain in economic benefits is to realize that society gets $750 worth of gasoline ($1.50 times 500 gallons) but it only costs $687.50. Thus the net gain is $62.50.

Because this change in output by this producer is so small relative to the market, there is no change in the price of gasoline to consumers. If there is no change in price, there is no change in consumer surplus and hence no net economic efficiency benefits to the consumer. In essence, this small additional amount of oil is added at the margin where price equals gross willingness to pay. Oil production will be reduced elsewhere, where production costs more than it now does here, by 500 gallons. We now turn to analyzing resource projects or policies that are so large in scope they do affect the market supply and hence price of the final good.

6.5 Benefits to Society with Price Changes

Consider a major water supply project or watershed yield enhancement project that would reduce a city's reliance on pumping ground water and replace it with surface water supplies from a proposed reservoir. Given the without-project supply curve (marginal cost curve) for water supply, the price of water is $15 per 1,000 gallons. Given this price, average consumption is 5000 gallons per person per month. The existing producer and consumer surplus are areas A and B in Figure 7-4. As Table 7-4 indicates, the consumer and producer surplus are $12.50 each in the without-project case.

The proposed project would reduce the cost of supplying water to the point where the supply curve shifts out to the Supply with Project in Figure 7-4. As this figure illustrates, the project would cut by 50% the cost of supplying the current 5000 gallons of water to city residents. With a fall in the price of water from $15 per 1000 gallons to $10, this cost saving is shared between consumers and producers. The exact division will depend on the relative slopes of the demand and supply curves. In our specific example the consumers gain$12.50 from the cost savings (area C1 in Figure 7-4) and $12.50 from the transfer of producer surplus to consumer surplus when the price of water falls from $15 to $10 (area B in Figure 7-4). The gain to the producer from the cost savings on the existing quantity of water supplied is area D1, or $37.50. However, the producer loses the producer surplus in area B ($12.50) when the price of water falls. Thus the net gain to the producer from the cost savings on the original 5,000 gallons of water is $25 ($37.50-$12.50). Notice that area B is not a net gain, but rather a transfer of producer surplus realized at the $15 price to consumer surplus when the price of water falls to $10.

In addition to the cost savings on the current quantity of water, the lowered price of water results in the quantity of water demanded increasing to 10,000 gallons. There is a consumer surplus gain from being

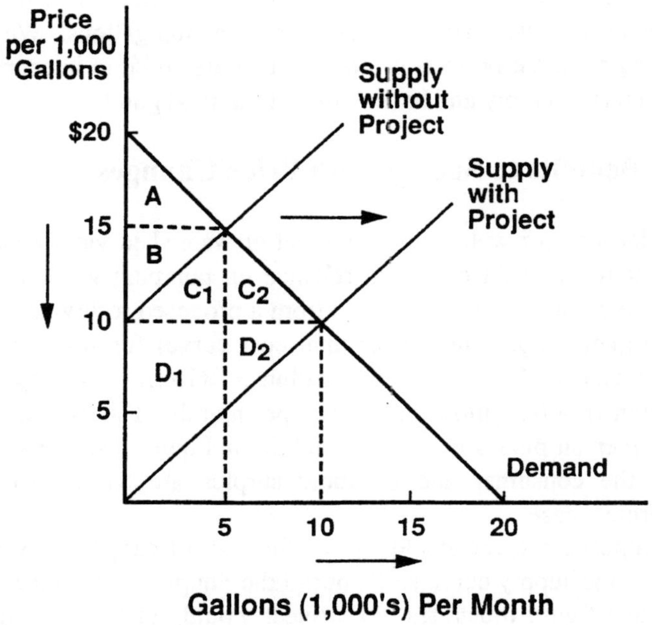

Figure 7-4 Change in Consumer and Producer Surplus with Price Changes

able to consume the additional water (via greener lawns, cleaner patios and cars, etc.) of area C2 or $12.50. The producer also gains a surplus from supplying the 5000 additional gallons of water. This gain is area D2, or $12.50.

Thus the total benefits of the project to city water users and producer is C1+C2+D1+D2, or $75 per month. This net gain can be calculated by subtracting the original producer and consumer surplus (B and A) from the new total consumer (A+B+C1+C2) and producer surplus (D1+D2). This is illustrated in Table 7-4. Specifically, the total consumer and producer surpluses with the project are $50 each, or $100 together. Given the without-project benefits of $25, the net gain is $75. Either way one measures the net gain, the project benefits are $75. This gain would be compared to the costs of constructing the project or performing the watershed management action to determine if the overall benefits ($75) outweigh the costs.

Table 7-4. Social Benefits from a Large Change in Supply

	Without Project	With Project	Net Gain
Producer Surplus			
Area	B	D1 + D2	
Dollars	$12.50	$50	$37.50
Consumer Surplus			
Area	A	C1 + C2 + B + A	
Dollars	$12.50	$50	$37.50
Total Benefits	$25	$100	$75

6.6 Consumer Benefits of Environmental Improvement

Demand curves reflect an inverse relationship between the quantity demanded and price, holding all other factors constant. One of the factors held constant is the quality of the good. It makes little sense to estimate a demand curve for cars that lumps together BMWs and Volkswagens. Much like demand shifts induced by changes in income, the demand curve will shift when the quality of the good changes.

This logic applies to environmental quality as well. If a town is currently using groundwater that is high in nitrates or bitter tasting because of residual pesticides or has a sulfur smell to it, then consumers would not be willing to pay as much for this water as they would for better quality water. If farmers reduce non-point source pollution or the city develops another source of higher water quality, the demand curve for drinking water will shift accordingly. As Maler (1974) first demonstrated, we can use the area between these shifted demand curves as a measure of the benefits to households for the change in environmental quality. A simple example will illustrate the point.

Figure 7-5 illustrates how the demand curve for drinking water would shift with an improvement in the safety or taste of the drinking water. At present the household's demand curve is given by D without. At the current price of water of $1 per 1,000 gallons, she consumes 6,000 gallons per month. The without-policy benefits are $9 per month.

If the water quality rises as a result of the pollution control policy or with the replacement supply of water, this increased utility from better tasting and safer drinking water shifts the demand curve outward to D with policy. This demand shift has two main effects. First, the household receives more utility on the current quantity of water consumed. The gain in consumer surplus on the existing 6,000 gallons is shown in Figure 7-5 as area A1, or $12 ($6-$4 times 6). In addition, the higher enjoyment and satisfaction when coupled with the original price of $1 per 1,000 gallons implies that consuming more municipal water is optimal. As illustrated in

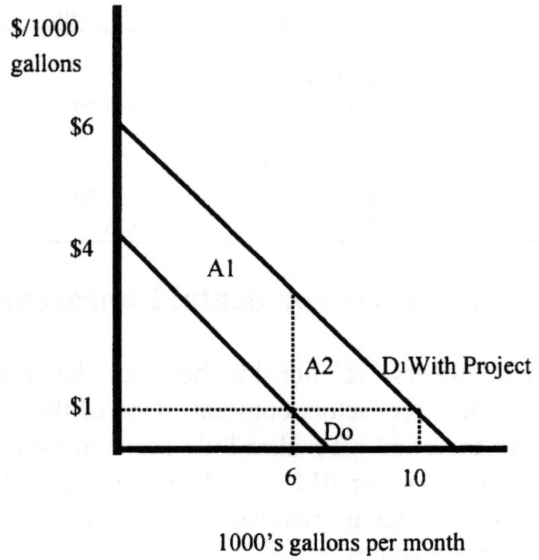

Figure 7-5 How Changes in Environmental Quality Shift Consumer Demand Curve for Water

Figure 7-5, the household increases the number of gallons consumed each month from 6,000 to 10,000. There is $4 of additional consumer surplus (area A2 in Figure 6-7) on this additional water consumed. Thus the total benefits to the typical household from the water quality enhancement is areas A1+A2, or $16.

We can calculate exactly the same answer for project benefits if we subtract the with-policy total consumer surplus from the without-policy consumer surplus. The with-policy consumer surplus in Figure 7-5 is $25 ($6-$1 times 10, divided by two). The without-policy consumer surplus is $9. The gain in monthly household benefits is $16, just as before.

6.7 Comparing the Value of Small Changes in Market Goods with Large Changes in Quantity or Quality of Nonmarketed Goods

Evaluating a typical public policy or regulation often involves comparing the value of a change in quantity in some marketed output, such as food, minerals, gasoline, paint, automobiles or paper, with a change in quantity or quality of a nonmarketed good (water quality, reduced cancer deaths from air pollution, endangered species protection). The marketed commodity is often traded in national or international markets. Typically, the domestic industry being considered for stricter regulation contributes only a small amount of output to this national market because of international production and trade in this commodity. For example, regulating non-point sources of pollution from feedlots in one state (e.g., Oklahoma) will often not affect the price of the marketed commodity, since it is produced in many other states and even imported from foreign countries. In essence, the demand curve for marketed outputs from this particular state is horizontal at the market price, just like the demand curve facing an individual gasoline station owner would be. Therefore, price can be used as a measure of gross willingness to pay for one more unit of commodity output. The net benefits of producing the marketed output from this tract of land would be determined by subtracting the costs of production of that marketed good from this gross willingness to pay or total revenue. Again this is analogous to the right side of Figure 7-4 for an individual gasoline producer. The benefits of commodity production are simply the producer surplus or change in net income realized. There is no consumer surplus to be added to the commodity production in this case, even though the first few units of the good traded in the market do provide a consumer surplus. The reason there is no consumer surplus in this case is that the change in supply is so small that it does not change the price. If there is no change in price of the good to consumers, there can be no change in consumer surplus. In essence, consumers continue to receive the same amount of consumer surplus they had without the policy. If the with- and without-policy consumer surplus is the same, there is no change in consumer surplus attributable to this policy.

However, many regulatory policies for improving air and water quality often make large or non-marginal changes in consumer surplus. This arises in part because most environmental quality (other than CO_2 or CFCs) is not a homogeneous product traded in national or international markets; the ability to trade widely and search for alternative producers is necessary for price (and therefore consumer surplus) to be unaffected. The non-marginal effect of regulatory policies on environmental quality also arises because there are few substitutes for improving local environmental quality. That is, if air quality in my city deteriorates because of urban sprawl-induced air pollution, I cannot find substitute air quality unless I move to another city. Thus my demand curve shifts inward, and I suffer a reduction in well-being, which would be measured by the change in consumer surplus to avoid that change (or willingness to accept compensation as discussed below).

Thus, there is no inconsistency here in terms of using market price to derive changes in producer surplus for market goods and then using consumer surplus for environmental quality. Both producer and consumer surplus are measures of willingness-to-pay. Price is a measure of gross willingness to pay at the margin for one more unit of the good. Thus, all resources are compared using the same conceptual measures of value, willingness to pay.

7. A NOTE ON WILLINGNESS TO PAY VERSUS WILLINGNESS TO ACCEPT

Since prices measure consumer's willingness to pay for one more unit of a good or service, it is natural to use willingness to pay (WTP) as a commensurate measure of value for market and non-market goods alike. This insures that all effects are treated equally. Economic theory suggests that WTP should be used only as a measure of gains. Since most policy evaluations are for improvements (after all, why should a policy make things worse?), WTP is frequently appropriate. However, to measure the dollar amount of losses, economic theory would suggest that a person's minimum willingness to accept (WTA) should be used. This is particularly true if the individual has a legal property right to the current level and the policy involves a reduction below what they currently have.

According to this same theory, the differences between WTP and WTA should not be large for goods or resources that make up a small percentage of a person's income and have many substitutes (Hanemann, 1991). If this were the typical case, then WTP, which is easier to measure, could be used as a good approximation of WTA for losses. Of course for many natural and environmental resources, such as water quality or native wildlife, there may not be many good substitutes, so WTA can exceed WTP by a substantial amount. Thus WTA may be the appropriate measure of losses if property rights are assigned to the current level and a reduction is being contemplated.

What becomes tricky for analysis is that often a policy will result in a reduction in a natural resource to individuals, but these individuals do not hold legal property rights to the current level of the resource. For example, I enjoy viewing the birds using a small wetland near where I live. A proposal to fill the wetland and make it a park would result in a loss in bird viewing. Should my loss be measured as my minimum WTA to accept the loss, or my maximum WTP to prevent the loss? Legally, I don't own the wetland, so strictly speaking the loss should not be measured by my willingness to accept compensation to allow it to be drained. However, one could presume that, since regulation of wetlands comes under the federal Clean Water Act, the public does hold some legal interest in the wetlands. Further, empirical evidence from numerous laboratory studies, many of which use real money, is that my minimum WTA will exceed my maximum WTP by a factor of 2 to 5 (Knestch and Sinden, 1984; Brown and Gregory, 1999:326). Thus, whether retaining the wetland or providing the park has the highest economic value may depend on which measure is used for valuing the wetlands.

The practical answer to this question often involves three key elements: (a) is there a public property right to the current level; (b) can WTA be reliably measured; and (c) what is the agency policy guidance for such analyses. Obtaining some legal opinion from lawyers in the agency enacting the policy may help clarify point (a). Even if legal opinion indicates that the public does have some limited vested right (along with the landowner or project proponent), can the analyst empirically estimate minimum WTA? While market data allowing estimation of demand curves often result in WTP estimates, there are computational algorithms (Vartia, 1983) and closed form solutions (Hausman, 1981) that can allow for estimation of WTA from estimated demand functions under certain circumstances (e.g., that income is a statistically significant variable). For some non-market resources that rely upon survey responses, the prospect of obtaining reliable estimates of minimum WTA are less rosy. Finally, the relevant regulatory agency may have made a high-level policy decision that WTP will be the standard measure of value. This has been the case for many U.S. agencies, such as the U.S. Army Corps of Engineers and U.S. Bureau of Reclamation, whose benefit-cost analyses are governed by the U.S. Water Resources Council Principles and Guidelines (1983).

If you are troubled by this potential ambiguity in the outcome of a benefit-cost analysis due to use of WTP vs. WTA, you can always perform a sensitivity analysis. As discussed elsewhere, this type of analysis calculates how sensitive the final answer is to uncertainty surrounding a key variable. One of those key variables could be whether to use WTP as a proxy for WTA. The net benefits of the alternatives could be calculated using both WTP and WTA for losses. Then the analyst, the decision maker, and the public can see whether the ranking of policy alternatives changes with the particular measure of losses used. If there is no change in ranking, then everyone can rest assured, the best policy alternative is robust to at least this change.

8. DIFFICULTIES IN APPLYING BCA IN ENVIRONMENTAL POLICY ANALYSIS

BCA can be a powerful analytical tool for comparing alternatives. Because it can be powerful, its use is often attacked. Sometimes those attacks are purely opportunistic, coming from those who prefer an alternative that does not have net benefits as high as the preferred alternative from the BCA. Some of the criticism of BCA is well deserved, though. Like any analytical (or other) tool, it has its valid uses, but it can also be misused. As a provider of BCA, you must be able to distinguish between valid criticism of BCA and attacks made because the attacker does not like the outcome. We group the criticisms into those that attack the conceptual foundations of BCA and those related to actual implementation of BCA.

8.1 Conceptual Difficulties with BCA

8.1.1 Anthropocentricity

BCA is explicitly an anthropocentric (people-centered) tool. The values that it includes are those that people have; it does not take into account intrinsic values associated with the natural world that are not derived from people. BCA is a people-based tool because people do it, and it is a decision-making tool for people. While it might not be impossible in theory for people to take into account other, non-human values, in practice there is no accepted way to measure those values independently of the values that people give them.

Thus, if people do not care about protection of a particular species, the benefit of its preservation, in the BCA framework, would be zero. However, it cannot be overemphasized that, if people value something, that something should be (and will be) included in the BCA. To the extent that people will speak for the intrinsic values of nature (and are willing to put their money where their mouths are), those values will appear in the calculation of nonmarket benefits.

8.1.2 Distributional Effects

A benefit-cost analysis is designed to determine the net benefits to society of an alternative. If the benefits of a policy exceed the costs, it is possible in principle for everyone to gain from that activity and for nobody to lose, through appropriate allocation of the benefits and costs across all of society. Of course, the fact that it is possible for everyone to gain does not mean that the necessary redistribution will in fact take place. Instead, a policy may benefit some groups at the expense of others.

For instance, protecting the northern spotted owl is estimated to produce benefits to people in the U.S. that exceed the costs of owl protection (specifically, the timber made unavailable for harvest by protecting the owl). However, for the timber workers who lost their jobs due to those reductions in timber harvest, protecting the spotted owl did not provide them with any benefits. As part of the plan to protect the spotted owl, retraining programs were provided to the displaced timber workers as partial compensation for their lost jobs. In this case, federal tax money (collected from the general population of the U.S., who benefited from protecting the owl) was used to reduce the pain to those who disproportionately bore the costs of owl protection. If the compensation had not taken place, there clearly would have been winners and losers. For those who lost, the fact that the net benefits were positive would almost certainly not have convinced them to support owl protection.

Typically, a BCA does *not* consider the distributional effects of a policy. Those distributional effects are critical for understanding who supports and who opposes different alternatives, and they often explain why some more efficient policies cannot be enacted. In addition, any decision-maker who does *not* consider the effects of a policy on interested groups is unlikely to stay in power for very long. Thus, distributional effects can and do frequently override the recommendations that come out of a BCA. However, if data and time permits, analyzing and displaying the distribution of benefits and costs among different groups can greatly improve the usefulness of the benefit-cost analysis to decision makers.

8.1.3 No Compromise: Trade-offs Among Non-Comparables

A benefit-cost analysis is based on the premise that a project has both benefits and costs, and that tradeoffs between these benefits and costs can and should be made. All of us would like a clean, healthy environment if it cost nothing to have one; we have environmental problems because there are costs involved in protecting the environment. In most cases, many people are willing to give up some valued goods to protect the environment, but there is a point where they are unwilling to pay more for additional environmental protection. For instance, many people are willing to pay a few hundred dollars more for a car with an engine designed to reduce air pollution emissions, but they may not be willing to give up their car completely and find an alternative means of transportation. In these situations, benefit-cost analysis can be an extremely useful tool for finding the right level of environmental protection.

In some environmental disputes, though, some people have a position staked out from which they are completely unwilling to move. For instance, a site might be sacred ground to a group of people; even if there is a billion-dollar gold deposit underneath it (and they would get the money), this group would accept no intrusion on the land. The implicit value of the site to them is infinite. In this case, a benefit-cost analysis will not influence their opinion; if the value of the site is infinite to them, its loss cannot be compensated. Indeed, it becomes conceptually difficult to conduct a BCA of developing a site if its value to the group truly is infinite, even if the value of other uses is also very high. For instance, suppose that the sacred site contains the only known deposit of a substance that is believed to have tremendous curative powers for some otherwise incurable human illnesses. Intrusion into the site would violate one group's fundamental beliefs, but those who will die without the substance would be willing to give up everything they have to get access to it. A benefit-cost analysis in this context would be virtually meaningless to either side. In these highly painful situations, the best role that an analyst can provide is to outline all the benefits and the costs, with or without price tags, and provide that information to a decision-maker. Even though benefit-cost analysis may not be directly useful, some systematic analysis may be possible using professionals trained in environmental ethics.

8.1.4 Compensation Paradoxes

Since a Potential Pareto Improvement does not require that compensation actually be paid, major concerns have been voiced about reliance on BCA for public decision making. The equity and distributional issues have been discussed above. Additional concern results from the possibility that implementing the new policy might change the relative prices of goods and income levels that influence individuals' willingness to pay. For example, a move from a situation where salmon are scarce and hydropower plentiful to an alternative situation where salmon are plentiful and hydropower is scarce could change relative prices of salmon and hydropower. In addition, the income of commercial anglers relative to owners and workers in businesses which use large amounts of electricity would likely change. If the effect of the policy was so widespread that relative prices or incomes substantially changed, then the desirability of the policy might change depending on whether we evaluate it using the original set of prices and incomes or the set of prices and incomes occurring with the change. This is what is referred to as a compensation paradox.

There are a couple of responses to this concern. First, most environmental policies make such small changes in the market supplies of goods or factors of production that relative prices or overall factor income shares rarely change. Thus the possibility of a compensation paradox is unlikely except for national policies having a substantial effect on multiple sectors of the economy. Second, in comparing an inefficient bundle of resource uses to an efficient bundle (i.e., a bundle that provides more of one resource and no less of another), the compensation paradox may not occur. The interested reader should see Just, Hueth and Schmitz (1982:38) for a more detailed discussion regarding the resolution of the compensation paradox.

8.1.5 Theory of Second Best

The purpose of BCA is to recommend projects, policies or programs that make people better off, i.e., represent an improvement in economic efficiency. In BCA we evaluate the economic efficiency of just the resources affected by the proposed policy. If the BCA indicates that we should allocate more water from agriculture to instream flow, then implementing such a recommendation would be thought to improve economic efficiency of the entire economy. That is, we have improved economic efficiency in use of recreation and fisheries and not negatively affected any other resources elsewhere in the economy (i.e., steel production, housing construction, clothing, etc.).

But the theory of second best (Lipsey and Lancaster, 1957) postulates that, if there are inefficiencies elsewhere in the economy, improving economic efficiency in one part of the economy will not necessarily increase social welfare. In other words, reducing economic inefficiency, as long as there are still some inefficiencies remaining, may not represent an improvement; there could be unintended consequences that reduce welfare. If this theory of second best is always relevant, then we could not reach any conclusion on the desirability of any single environmental policy based on the information contained in a BCA as long as there were other industries or resources uses that remained economically inefficient.

Fortunately, re-evaluation of the theory of second best has shown that we need not be so pessimistic about recommending economically efficient uses of natural resources in an otherwise inefficient world. First, Mishan (1981:293) indicates that, any time we can produce the same bundle and level of output for less cost, there is an unambiguous gain to society. Secondly, as Davis and Whinston (1965) have shown, if there is separability or independence between economically inefficient sectors of the economy and the ones under study in the BCA, we may be able to make unambiguous statements about improvements in economic efficiency. For many small scale policy analyses there is often little connection between a particular set of firms in a given geographic location and numerous other sectors of the economy (i.e., clothing, steel production or housing construction, etc.). Thus, concerns over second best need not preclude us from concluding that increased efficiency in environmental issues will improve social well being.

8.2 Practical Difficulties with Actual Implementation of BCA

8.2.1 Incomplete BCA

Probably the most common problem with BCAs is that they frequently exclude parts of the problem due to lack of data or due to lack of an ability to quantify one aspect into dollars. This is not a problem with the conceptual underpinnings of BCA; rather, it is a problem with its implementation, and a far too common one. For instance, Federal Energy Regulatory Commissions economic analyses of relicensing of private hydroelectric projects include the value of hydroelectricity but do not always include the dollar value of recreational fishing and rafting. A benefit-cost analysis, if it does not include all benefits and costs, will produce a biased answer.

If a BCA is incomplete, it is up to the analyst to ensure that the client for the BCA knows its limitations. Typically the marketable aspects of the project are much easier to monetize than the non-marketed and environmental effects. If the environmental effects cannot be monetized, at the least they need to be discussed as thoroughly as possible. For instance, if the study budget does not allow for the analyst to quantify in dollars the benefits of preserving a wetland area, the analyst should identify and (to the extent feasible) quantify in other units the benefits of preservation. For example, how many endangered or threatened species depend on the wetland? Or how many visitor days of recreational use does it get? How widespread in the public is interest in preserving the area? Not providing information as complete as possible about the effects not included in the BCA can lead to bad decisions, especially regarding environmental impacts.

Even if not all effects can be quantified in dollars, an incomplete BCA may still be useful in comparing among alternatives. Suppose that the unquantified effects do not change between the alternatives. For instance, suppose alternatives A and B will provide the same level of wilderness preservation, but A produces more timber, while B produces more minerals.

In this case, using BCA to compare between mining and timber will not affect wilderness, and its exclusion from the BCA will not affect the identification of the optimal extractive use. If alternative C would expand wilderness and reduce timber harvesting, though, it cannot be compared to alternative A. A second case was illustrated in the grazing versus wetland example in Table 7-3. Here only the waterfowl hunting benefits associated with the wetland were quantified, and not any of the other ecosystem services of the wetland, such as groundwater recharge. Still, having quantified the livestock and hunting benefits and costs, we were able to ask whether the unmeasured benefits of the wetland were greater than the $10,000 net gain of grazing over the hunting benefits. In this way even an incomplete benefit-cost analysis can help to frame the key trade-off in a way that focuses the decision.

8.2.2 Decision Tool vs. Decision Rule

The use of benefit-cost analyses has often been criticized because some economists and decision-makers view the results of a BCA as *the* decision rule. In other words, some believe that BCA, by determining the alternative with the greatest net economic benefits, automatically recommends that alternative for adoption over all others. In this view the BCA "makes the decision." Even when those conducting the BCA do not adopt that view, many believe that the intention of a BCA is to provide a final decision. If the BCA supports an alternative advocated by one group, that group may argue in favor of the use of BCA for a decision rule, while those opposed to that alternative will argue either that BCA is an inappropriate method for decision or that the BCA was conducted inappropriately. Thus, sometimes a debate over the recommendation arising from a BCA turns into a debate over the legitimacy of BCA itself.

While the mechanics of performing BCA are the same regardless of whether it is considered an informational input to decision-making or a decision rule, the use of the BCA is quite different with the two views. There will always be social and environmental concerns that will not be adequately quantified into dollars or adequately represented through a single number, making even the best BCA incomplete as a description of the totality of effects of some policy on all members of society. Therefore, expecting the BCA to make the policy decision is only appropriate if the law or regulation the analyst is operating under explicitly assigns primary value to economic efficiency factors. Otherwise, in the spirit of this book, BCA provides just one viewpoint, albeit often an important one.

In sum, there are a number of valid objections to the use of BCA. At the same time, it is hard to think of circumstances where a decision-maker would *not* like to know the net impacts of an alternative on society. Does it increase net societal well-being, so that it is possible to make nobody worse off and at least some better off? If not, on what basis can the alternative be supported? Are those who gain from the project so much more important that the net losses to everyone else are justified? Though using a BCA as the sole decision rule among a set of alternatives is likely to come under criticism, not using it as an input to the decision process is likely to come under objections at least as strong.

Chapter 8

Discounting Benefits and Costs Over Time

Many new policies are intended to have long-lasting effects. They may require firms to invest in long-lived wastewater treatment plants, or influence the permanent conversion of land from one use to another. Federal relicensing of hydropower dams is often for a 30-40 year period. Decisions to allow off-shore oil leasing will have effects that last for decades as the oil is pumped. Construction of highways and light rail systems are not only long lived, but also have large up-front costs, with benefits that are received for many years to follow.

Can we simply add up the dollar benefits in the first year with the dollar benefits in each of the following years? Is a dollar's worth of benefits 20 years from now equal to a dollar's worth of costs incurred to build the project? Comparing a dollar's worth of costs today to a dollar's worth of benefits in 20 years implies that people are equally willing to trade a dollar today for a dollar in 20 years. We know that is incorrect. You might suspect it is incorrect due to inflation, but that is not the only problem here. While inflation is certainly an issue, the benefits and costs of policy actions are normally measured in constant dollars, that is, adjusted for inflation. Therefore, the main issue is whether a dollar of highway costs today is equal to an inflation-proofed or real dollar of highway benefits in the future. Put another way, would you give me a dollar today if I gave you a written guarantee to give you a dollar (fully indexed up in value for inflation) 20 years from now?

Chances are you would not treat a dollar today and an inflation-indexed dollar 20 years from now as equivalent. There are two reasons we find people unwilling to trade a dollar now for an inflation-indexed dollar 20 years from now. First, people have a time preference for consumption.

Other things equal, people prefer their enjoyment or benefits now rather than later. This explains why people pay an interest premium to borrow money from banks or on their credit cards to consume now and pay later. Secondly, if people are to forego consumption benefits today, they can invest the money in some productive enterprise that will yield a net return. That is, people deposit money in a bank, or loan money to a business, that will pay them a positive rate of interest because the money is being put to a productive use. The result is that a dollar in 20 years is worth less than a dollar now, because you can use the current money either to increase your current consumption, or you can invest it and earn interest.

1. DISCOUNTING PROCESS

In some ways, the interest rate reflects both humans' 'time preference' for money and the rate of return that could be earned by investing this money. Most people are familiar with the concept of compound interest rates, whereby money put in the bank today will grow at some compound rate of interest to a larger sum of money in the future. For benefit-cost analysis, where we wish to take a future flow of monetary benefits and figure out the value in the present period, we use a process that is just the reverse of compounding, called discounting.

Discounting takes those future (real or inflation-indexed) dollar benefits and costs and computes their present worth, or present value. That is, exactly how much benefit does $10 worth of hydroelectricity twenty years from now provide society today?

An example may help illustrate this process. If the interest rate is 7.125% this implies that $1 of benefits 10 years from now is worth 50 cents to us today. How did we figure that? Well, if I put the 50 cents in the bank today at 7.125%, I could take out $1 in ten years. Thus, at 7.125% interest, 50 cents today and $1 in ten years are equivalent in terms of their present worth or value to me today. In some sense that is all there is to it. If we know what interest rate people use to be indifferent between present and future consumption (or the rate of return that could be earned on funds invested), then we can convert any future benefit into an equivalent present value. We can do this for the entire stream of benefits and costs over the life of some public project or policy and determine if the present value of the benefits is worth the present value of the costs.

The formula for calculating the present value of any dollar amount ($\$B_t$) to be paid or received t years from now is:

$$\text{(8-1)} \quad \text{PV of } \$B_t = \frac{\$B_t}{(1+r)^t}$$

Most calculators or spreadsheet programs have financial functions that allow the calculation of present values for a series of benefits or costs over time. For example, at 10% interest, the present value at year zero of $500 a year in benefits and an initial investment cost of $2000 with annual operation and maintenance costs of $200 a year would be calculated as shown in Table 8-1.

Table 8-1. Comparison of Undiscounted and Discounted Benefits and Costs at 10%

Year	Discount Factor	Benefits		Costs		Net Benefits
		Nominal	Discounted	Nominal	Discounted	
1	0.9090909	$500	$454.55	2000	$1818.18	-1363.63
2	0.8264463	$500	$413.22	200	$165.28	247.94
3	0.7513148	$500	$375.66	200	$150.26	225.40
4	0.6830135	$500	$341.51	200	$136.60	204.91
5	0.6209213	$500	$310.46	200	$124.18	186.28
6	0.5644739	$500	$282.24	200	$112.89	169.34
7	0.5131581	$500	$256.58	200	$102.63	153.95
8	0.4665074	$500	$233.25	200	$93.30	139.95
9	0.4240976	$500	$212.00	200	$84.80	127.20
10	0.3855433	$500	$192.77	200	$77.11	115.66
Discounted Sum			$3,072		$2,865	$207
Net Present Value						$207

The discount factor here is $1/(1+r)^t$, where r is the interest rate, and t is the number of years. It is multiplied by the nominal value to give the present discounted value.

The net present value can be computed by either subtracting the present value of benefits from present value of costs, or discounting net benefits (undiscounted benefits minus costs). As can be seen in this example, either approach to computing the net present value results in the same answer, $207. Also note the error in calculating net benefits that would have resulted had the undiscounted benefits of $500 a year times 10 years ($5000) been subtracted from the costs of $3800. This would incorrectly conclude the project is worth $1200, because it ignores the differential timing of costs compared to benefits.

Figure 8-1 provides a graph of the distribution of present value of one dollar received in each of 20 years at a 4%, 7% and 10% discount rates. Note the rapid drop off at 10% as compared to 4%. As will be discussed more in detail below, the 7% rate is what most federal agencies use.

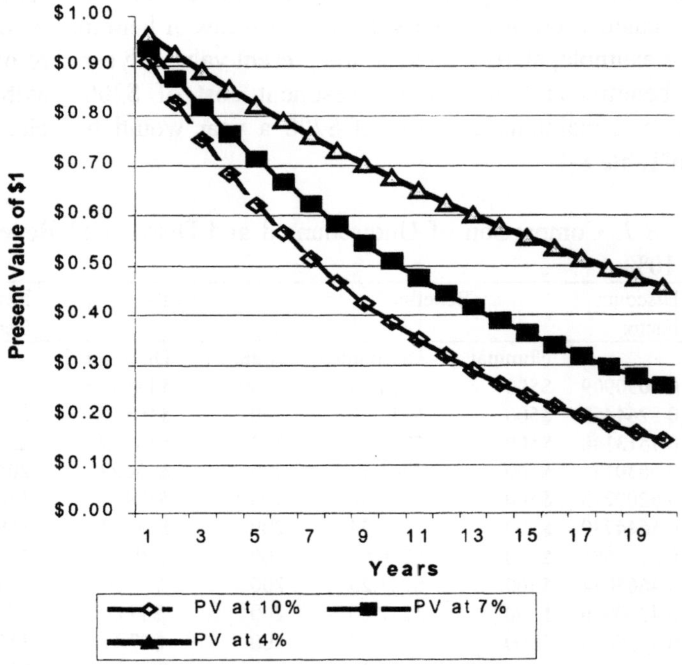

Figure 8-1
Comparison of Present Value (PV) of $1 at Three Discount Rates

2. SELECTING THE INTEREST OR DISCOUNT RATE

The selection of the appropriate discount rate is of both philosophical and practical importance. There is an almost ethical or philosophical statement in using high discount rates: the present matters a great deal more than the future to us. If everyone was equally happy to have a dollar's worth of consumption today as in ten years, this would imply a zero discount rate. However, since money invested in productive enterprises today has the potential to yield a net return in the future, even an indifference of present for future consumption is not enough to yield a zero discount rate to society. However, Lind (1990: 21) cautions that discounting is most defensible over time periods not exceeding a typical person's lifetime (about 70 years in most developed countries).

2.1 Components of Observed Interest Rates

Economic theory provides clear guidance on selection of discount rates in an idealized economy of zero financial transactions costs, zero risk, and no taxes (we said it was idealized!). In this case the time-preference value of money (trade-off of present for future consumption) equals the rate of return that could be earned by investing this money. Once real world distortions are added in, these two measures of the discount rate diverge. To understand this, it is important to realize that the observed market rate of interest reflects four factors: (1) the real or true rate of discount; (2) a percentage return to compensate for inflation; (3) a percentage return to reflect risk that the borrower might not repay; and (4) tax rates. At its simplest, if inflation is expected to be 5% over the future, there is a 2% risk premium, and the real rate of return is 6%, then the market interest rate before taxes would be 13%. If the tax rate is 25%, the after tax interest received is only 9.75% (13% times 0.75).

In general we would want to use the full 13%, since taxes are just a transfer of the rate of return from the individual to the government (and hence society), and the taxed part of the rate of return is actually not lost. If we keep everything in real or year zero prices, then we would not include anticipated inflation, so a real discount rate would be just 8% (6% + 2%). Some economists have argued that the federal government and many state governments are so large that they are able to pool risks across hundreds of projects or policies being implemented. As such, the risk premium should be zero. In this case the real discount rate for public investments would be 6%, just the real rate of return. We now turn to the economic theory underlying what real rate of return should be used.

2.2 Theories for Determining the Discount Rate

There are several competing theories on the determination of what the discount rate should be given the many real world distortions. One of the most prominent theories dates back to a macroeconomic view of a national economy as a closed economy (i.e., an economy where imports and exports of both goods and capital are nonexistent or minimal). A simple macroeconomic model of such an economy requires that the sum of consumers' savings and taxes equal government spending and private investment. In this economy, if the government wishes to spend more money to invest in a public project, one of the other three elements must change accordingly. If the government borrows the money for the project, this simple model argues that private investment must fall by the amount of money the government must borrow (i.e., complete crowding out of private investment). If the government raises taxes to finance the project, then consumption and savings will fall by the amount of the tax increase.

Given these relationships between government financing of a public project and displacement of economic activity elsewhere in the economy, one group of economists argue that the discount rate should emphasize the opportunity costs in the private sector (foregone investment, consumption and savings) that results from diverting the funds to the public sector (Baumol, 1978). The opportunity costs in the private sector are made up of two components: the rate of return on displaced private investments, and the value of foregone consumption. The rate of return on investments can be obtained from financial reports of companies. The value of foregone consumption (for small changes in consumption) can be inferred from the interest rate people would earn on savings. That is, as consumers decide whether to consume their last dollar or save their last dollar, they compare the value of consumption to the interest rate they would earn on savings. The overall social discount rate is the weighted sum of the private investment rate of return and the interest rate on savings. The weight is determined by how much of the monies to be transferred to the public sector come from displaced private investment and how much comes from displaced savings/consumption. This ultimately depends on how the government raises the money for the investment project. If special financing legislation, dedicated trust fund dollars, or specific excise taxes are used, then, in principle, a project- or policy-specific discount rate could be calculated. In practice this is unusual; the weights would come from the proportion of the government's budget that comes from individual income taxes, corporate income taxes, other business taxes, and other consumer taxes received by the government.

Today, however, most national economies are part of an open world economy, where substantial exports and imports (including foreign capital) make the closed economy view somewhat misleading (Feldstein, 1985). No longer must private investment be reduced dollar per dollar with the increase in a deficit-financed public project. Directly or indirectly, foreign owners of capital can meet the new increased demand for funds arising from the government's project. In an open world economy there is very little crowding out of domestic private investment (Lind, 1990: 15). As such, Lind's recent reassessment of determination of the discount rate recommends the appropriate government discount rate (in open world economies) be the interest rate the government must pay when it borrows (1990: 25).

Of course an interesting wrinkle to this occurs when the government policy does not require the government to make investments, but instead requires the private sector to make investments. For example, workplace-based regulation that requires firms to invest in reducing worker exposure to carcinogens may require the firm to spend millions of dollars on new filtration systems. In this case, it is industry's opportunity cost of capital or their cost of borrowing, not the government's, that may be relevant for determining the present value of the costs. In this case, risk would be a real cost, but taxes are still a transfer.

Another set of arguments (e.g., Arrow, 1965) stresses the "social time preference" in determining the discount rate. In particular, these arguments start by viewing the discount rate as a measure of people's preferences for present versus future consumption. They then introduce a public goods or beneficial externality argument. This argument is that individuals care more about future generations than their private savings decisions might indicate. Acting individually, their individual savings decisions have little effect on the welfare of future generations. However, collectively society does care, and if all people would save more (and consume less) for future generations, we would all be better off. This theory is an argument for discount rates lower than the individual savings rate to encourage investing for future generations.

Recently, Weitzman (1998) has addressed the issue of selecting discount rates for environmental projects and policies whose effects may last for centuries. Examples of such policies include global climate change, disposal of radioactive waste, groundwater pollution, and species extinction. Normal exponential discounting as given in equation (8-1) with discount rates typically used by Federal agencies would render even large benefits received or costs incurred 200 years from now as trivial. For example, at 7%, a dollar 200 years from now would be one-millionth of a dollar today. Thus, foregoing $2 in benefits today to prevent a $1 million dollar cost 200 years from now would not pass a benefit-cost test. As Weitzman points out, this makes even most economists feel uneasy, let alone those who feel a strong moral obligation not to impose costs on unrepresented future generations. But what's a policy analyst to do? Weitzman suggests a potential solution for policies with such far distant effects. Using the view that discount rates should reflect the long-run productivity of capital, he suggest there is little reason to treat the discount rate as a fixed constant over several centuries. There is a great deal of uncertainty over technological progress and a host of factors influencing the productivity of capital over the next 2-3 centuries. What we really have is a distribution of long-run productivities of capital and associated distribution of discount rates. As Weiztman then demonstrates, it is only the realization of the lowest discount rates that really have any impact on current decision-making (1998:205). The high discount rates result in such high discount factors that those future benefits and costs do not matter today. This can be made more intuitive by looking back at Figure 8-1 and comparing the far more rapid rate of decline of the present value of $1 with the 7% as compared to 4% discount rate. Weitzman (1998:207) concludes, "Thus, the paper is suggesting at least the possibility that it may be essential to incorporate declining discount rates into any benefit-cost methodology for evaluating long-term environmental projects." While this does not provide a simple solution, it does suggest that a sensitivity analysis of the net present value of environmental policies with long term effects be performed. For example, the analyst could have discount rates declining each decade from their current level toward something approaching zero by the end of the next century.

Weitzman (2001) has recently gone further to indicate that, rather than relying on a single discount rate, a discount function should be used, with a discount rate of 4% for projects in the immediate future, 3% for near term (6-25 years), 2% for projects of 26 to 75 years, and finally 1% for effects of up to 300 years in duration.

3. WHAT FEDERAL AGENCIES USE

It would often be impractical and socially inefficient for each federal agency to compute its own discount rates on each and every public project. As noted below, it would also lead to overall sub-optimality of public investment. Instead, Federal policy sets discount rates that agencies must use. Much of the Federal government uses a real rate of interest of 7% set by Office of Management and Budget (OMB, 1992). OMB established this rate using the theory that the appropriate discount rate relates to the opportunity cost view, in particular, the foregone pre-tax rate of return on investments in the private sector. Since this discount rate is a real rate, it does not contain any accounting for inflation. Therefore the 7% rate maybe somewhat high, although interest rates in the private sector may be higher.

Because Congress set a separate formula for discount rates for water resources, OMB guidance specifically exempts the water-related Federal agencies (U.S. Army Corps of Engineers, U.S. Bureau of Reclamation, Natural Resources Conservation Service) from its guidance. The water resources agencies' discount rate reflects the interest rate cost to the federal government of borrowing money: specifically, the interest rate on long-term Treasury Bonds. This rate has been in the mid 7% range during the 1990s. This is a tax free rate, since individuals do not have to pay Federal income taxes on it. It is a risk free rate as well. This is a nominal rate of interest since it includes a premium for inflation. However, since most Federal agency benefit-cost analyses are not allowed to adjust future benefits and costs for inflation (that is, they do them in real dollars or project period values), it is incorrect to include inflation in the discount rate. This is an example of the "Congress told us to do it that way" dilemma in which Federal agencies often find themselves. Nonetheless, this U.S. Water Resources Council discount rate approach does accord (with the exception of being a nominal rate) with Lind's theoretical development that in an open world economy, the government borrowing rate is the appropriate discount rate as well.

Besides water resource agencies, a few other entities use different discount rates. For example, the U.S. Forest Service uses a 4% real discount rate. This rate was arrived at after analyzing the rate of returns on capital and corporate bond rates. This rate correctly considers taxes (i.e., it is a before tax rate) but excludes risk (Row, Kaiser and Session, 1981). The National Oceanic and Atmospheric Administration (NOAA) uses a 3% real rate for discounting damages from oil spills (Julius, 1999). Their justification for the 3% real rate is that it is the best estimate of social time preference of consumers. In particular, 3% has been the real after-tax rate of return to consumers (savers) in adjusting their time profile of income and consumption. One reason this rate is much lower than OMB's is that it uses the after tax (net) return rather than OMB's pretax return to businesses. However, NOAA indicates the 3% rate of return is also consistent with recent short term (3 months) borrowing costs to the Treasury. This is the basis for Department of Interior's recommendation to use 3% as well. Even OMB acknowledges that 3% is the real interest rate on Treasury notes and bonds. OMB allows agencies to use such an interest rate when discounting cost savings to the government under cost-effectiveness analysis. EPA (U.S. EPA, 1998b) recently argued for use of the 3% rate of time preference in analyzing lead abatement in homes since its hazard standard mainly affected people as consumers. Given minimal foregone investment effect of the regulation, the 3% social time preference of consumers was deemed more appropriate than the 7% opportunity cost of foregone investment.

While the 3% discount rate is probably closer to the social time preference rate, there is an overall sub-optimality in comparing the economic efficiency of public projects calculated using different discount rates. For example, the Department of Transportation might reject a regulation that would only yield a 6% rate of return because this return is below the OMB discount rate of 7% (i.e., what the funds are assumed to yield in competing investments). The NOAA might accept a restoration project that yields a rate of return of 4% because it exceeds their 3% discount rate. What of course happens here is the nation is foregoing a transportation regulation offering a 6% rate of return for one that offers only a 4% rate of return. Thus, it is at least as important to have a consistent discount rate across all public agencies as it is to have one that is theoretically correct. While there does seem to be movement toward a common real rate of 7% in the Federal government, this process needs to be accelerated so that all Federal agencies (transportation, occupational safety, water resources, etc.) use the same rate unless there is a very strong rationale for another rate.

We will now turn to a discussion of how to use discounted benefits and costs in determining an aggregate measure of the economic efficiency effects of an action.

4. ALTERNATIVE BENEFIT-COST DECISION CRITERIA

Once the analyst has measured the benefits and costs over the life of the project for each affected resource output, the next step is to summarize this information in a bottom line. For the information to be most useful for decision making, it is helpful to compare the sum of the discounted benefits produced by the alternative to the discounted costs incurred by the alternative.

The criteria and summary measures which are most defensible and commonly used include: (1) Net Present Value, (2) Benefit Cost Ratio, and (3) Internal Rate of Return. We will discuss each of these individually and then discuss under what conditions they yield exactly the same ranking of alternative policies or projects.

4.1 Net Present Value (NPV)

As the name implies, Net Present Value (NPV) is the present value of the benefits minus the present value of the costs. The difference is the net gain adjusted for the timing of benefits and costs. The units of measurement are present worth of dollars in the base year in which all of the benefits and costs are figured.

The definition of NPV can be understood by reference to equation 8-2:

$$(8\text{-}2)\ \text{NPV} = \sum_{t=0}^{T} \frac{B_t - C_t}{(1+r)^t}$$

Where :
B_t = Benefits in time period t
C_t = Costs in time period t
r = interest rate
t = time period

As equation 8-2 illustrates, the benefits in each year are subtracted from the costs in each year, and this difference is discounted back to the present time period (t=0). If the resource management action's present value of benefits exceed the present value of costs, then NPV will be greater than zero, i.e., positive. In general, the larger the NPV is, the more net benefits society realizes from the resource management action.

If the analyst is using NPV as a decision rule, then only projects with NPV greater than zero are accepted as economically efficient. Projects with negative NPVs mean society gives up more than it gets over the life of the resource management action. NPV is a discounted measure of whether the resource management action represents a Potential Pareto Improvement over the life of the project. Taking benefits as measures of gains and costs as a measure of losses, if NPV is greater than zero, then the gainers over the life of the project could compensate the losers over the life of the project and still have some benefit remaining.

4.2 Benefit-Cost Ratio (BCR)

The Benefit Cost Ratio (BCR) takes the same information used in equation 6-2 but expresses it as a ratio. Specifically, BCR equals the present value of the benefits divided by the present value of the costs.

Equation 8-3 presents this relationship:

$$(8\text{-}3)\ BCR = \frac{\sum_{t=0}^{T} \frac{B_t}{(1+r)^t}}{\sum_{t=0}^{T} \frac{C_t}{(1+r)^t}}$$

Since BCR is by definition a ratio, it will always be a positive number. If the discounted benefits exceed the costs, then BCR will be greater than one. Thus a BCR of 3:1 means that society receives three dollars worth of benefits for each dollar of cost. Thus BCR is an average return yielded by the project or resource management action.

While it is sometimes believed that a higher BCR is better, a very high benefit cost ratio may mean the project is too small, and its size and scope should be expanded. If BCR is being used as a decision rule, then only projects or resource management actions that have BCR greater than one would be considered for adoption.

4.3 Internal Rate of Return (IRR)

The Internal Rate of Return (IRR) is the percentage return on money invested in the project that equates the present value of the benefits with the present value of the costs. In essence, IRR is the earning power or interest earned when the money is invested in the project under study. Mathematically, IRR is that interest rate which results in the NPV of the project equaling zero. Thus the IRR formula solves for r in Equation 8-4 such that NPV = 0 or:

$$(8\text{-}4)\ NPV = \sum_{t=0}^{T} \frac{B_t - C_t}{(1+r)^t} = 0$$

Unlike the BCR or NPV, which can easily be computed on even a simple calculator, determining the IRR involves a series of relatively complex calculations. As such it is best performed using financial calculators that are programmed to calculate IRR, or a computer software program designed to calculate IRR. By itself, this added computational burden is not a valid reason for reluctance to use IRR (but there may be other objections that will be explained below).

To use the calculated IRR to determine the economic efficiency of a project involves a simple comparison. The calculated IRR is compared to the interest rate at which the agency or society can borrow money. As will be recalled from our earlier discussion, one view of the interest rate is the opportunity cost of foregone rates of return in society's next best investment. Thus the IRR decision rule is to recommend any project which generates an Internal Rate of Return greater than the interest rate.

4.4 Does It Matter Which Discounted Measure Is Used?

A comparison of the three formulas for NPV, BCR and IRR indicates that all three use the same information. Therefore, one would expect these three measures would yield the same conclusions on which projects should be adopted and which should be rejected. However, this is only true under certain conditions.

The conditions under which all three measures will yield the same conclusions regarding the desirability of projects are: (1) Either there are no budget limits that preclude adopting all economically efficient projects that pass their respective decision rule OR the size of projects in terms of investment costs are similar. (2) The projects or policies are not mutually exclusive. That is, each project could be implemented without precluding implementation of another. (3) Either the projects being compared have a similar length of time (i.e., economic life) OR the reinvestment opportunities yield the same returns as the original project.

If these conditions are not met, what appears to be the best project alternative will vary depending on which discounted measure (NPV, BCR, IRR) is used to evaluate the alternatives. For example, Table 8-2 involves violation of both aspects of condition 1 and part of condition 3. First, the projects have substantially different investment requirements. If we add to this the budget constraint that society only has $10 million to invest, then we must choose between the projects rather than adopting both (which is warranted since both are economically efficient regardless of whether one uses NPV or BCR). In addition, we are unable to find reinvestment opportunities that would yield the same return as we get from doing several project As (thus the default assumption is that the remaining $8.3 million of the original $10 million not invested under alternative A would earn just the discount rate).

Under these circumstances, which project is best to select depends on whether we rank by NPV (choose Project B) versus rank by BCR (choose Project A).

Table 8-2. Comparison of Benefit-Cost Ratio to Net Present Value (Millions of Dollars)

Project	PV of Benefits	PV of Costs	NPV	BCR
A	4	1.7	2.3	2.35
B	16	10	6	1.6

An example of this type of situation is the decision whether to build a light rail or to add more buses. The light rail will have much higher investment costs than adding more buses, yielding a lower BCR, but higher NPV. Since we are likely to find situations where governments have very limited budgets to devote to public projects, or nature makes the projects or policies mutually exclusive, it is important to know the advantages of each of the three discounted measures of project worth.

4.5 Advantages of NPV as a Theoretically Correct Measure of Net Benefits

We can illustrate the merits of NPV relative to BCR by reference to Table 8-2. When conditions 1-3 above do not hold, the choice between Plan A and Plan B becomes clear: does society prefer $6 million in net benefits or $2.3 million in net benefits? If all of the opportunity costs are correctly measured in the present value of costs, and the discount rate accurately reflects the foregone investment opportunities, then Plan B is preferred. People prefer more ($6 million) to less ($2.3 million) if these are mutually exclusive projects and we cannot invest in several identical project As. Our theory of what is an improvement in well-being is written in terms of maximizing the total amount of the net gain, not in terms of maximizing the rate of return. A very high rate of return for which we are restricted to collecting on only a $1 investment is not attractive if it means giving up a larger total return on some other investment. (Note that if the remaining $8.3 million not invested in project A just earns the discount rate, then this would pull down the BCR of project A to 1.23, again demonstrating the superiority of project B).

Another way to see that project B is superior to project A is to compare the marginal benefits and the marginal costs of the additional expense involved in B. While Project B includes $8.3 million more in costs, it provides $12 million more in benefits, justifying the additional expenditure.

As Boadway (1979) has shown, maximizing NPV is a necessary condition for a project or policy to result in a Potential Pareto Improvement.

Maximizing the BCR is not strictly related to attainment of Potential Pareto Improvement, since the BCR is the average rate of return. A high average rate of return tells us little about whether the total benefits are maximized. A high average return can be consistent with a project which is not large enough (marginal benefit from expansion greater than marginal cost of expansion, i.e., MB > MC). To attain the maximum gain, marginal cost must equal the marginal benefit (MB = MC). Thus NPV is a useful criterion for determining the optimal scale of a resource management action as well as timing of implementation. The BCR is an ambiguous indicator of the timing or the scale.

Additionally, the actual value of the BCR can be influenced by how costs and benefits are defined. For instance, suppose that a project with costs of $5 million will increase developed recreation by a value of $20 million but reduce dispersed recreation by $10 million. Is the reduction in dispersed recreation a reduction in benefits, or an increase in costs? In the former case, the BCR will be ($20 - $10 million)/($5 million) = 2; in the latter case, the BCR will be ($20 million)/($10 + 5 million) = 1.33. In either case, the NPV = $5 million. The BCR will always be greater than 1 if the NPV is positive (and less than 1 if the NPV is negative), but the exact value of the ratio depends on whether foregone benefits are treated as a cost and placed in the denominator or treated as a negative benefit in the numerator.

4.6 Advantages of BCR

There are, however, a few merits to use of the BCR. Some authors claim (Sassone and Schaffer, 1978:21) that, if one must rank many projects within a budget constraint, use of BCR and calculating incremental BCRs is the appropriate procedure. In addition, the BCR is easily understood by the public because it is expressed as dollars of benefit per dollar of cost. Thus it has a nice intuitive appeal. Lastly, in comparing projects where one does not want the scale of the project to influence opinions about desirability, the BCR is unaffected by the scale of the projects.

4.7 Merits of IRR as a Measure of Discounted Net Benefits

So far we have focused on the relative merits of NPV and BCR. However, the IRR has a similar set of advantages and disadvantages. Because the IRR can be expressed as a return per dollar of investment, it shares with the BCR its ease of understanding by the public and being unaffected by the scale of the project. However, since it also shares BCR's drawback of using a rate of return rather than total return as the decision rule, it too suffers from being a poor indicator of which project provides the greatest total gain to society. What is worse about IRR is the possibility that there exist several interest rates which set NPV = 0. Depending on the pattern of costs and benefits over time, there may be two or three or more interest rates (r) in equation 8-4 that are possible answers. Unfortunately, the analyst is likely to be completely unaware of exactly which interest rate has been solved for.

However, the IRR has one advantage not shared by BCR and NPV: the ability to make comparisons between alternative projects or policies without first having to specify an exact discount rate. In essence, the IRR formula solves for the discount rate (i.e., IRR) which would just make the present value of benefits equal to the present value of costs. If the actual discount or interest rate is lower than the internal rate of return, then this project will provide net benefits in excess of its opportunity cost. Hence this project or policy option would be acceptable on economic efficiency grounds. However, in many instances an agency or decision-maker need not state the exact discount rate it is using. As long as the project or policy generates a "reasonable" return (given the implicit discount rate the agency is using), the project can be accepted without ever having to decide the exact discount rate to use. The practical advantage of such an approach should be obvious given the earlier discussion about the difficulty in getting universal agreement on what should be the appropriate discount rate.

However, as also discussed earlier, most agencies have their discount rate specified, so this feature of IRR is not always compelling. In addition, many agencies have their decision criteria (i.e., NPV, BCR and IRR) specified as well. Agencies such as the Forest Service, Army Corps of Engineers, and USDA Natural Resources Conservation Service generally rely on use of NPV, although BCRs are also computed as supplemental information.

In general, use of NPV as the Benefit-Cost Analysis decision criteria has a great deal to offer. It is the most theoretically correct measure in the sense of being most consistent with the conceptual foundation of BCA. NPV does not suffer the multiple solutions problem that can potentially plague IRR calculations.

5. SENSITIVITY ANALYSIS TO ADDRESS UNCERTAINTY

There will always be some uncertainty regarding the exact amounts of future benefits and costs associated with a project or policy. There are frequently several sources of uncertainty. These include biological, medical, ecological uncertainty about the effects of the policy on species or human health, uncertainty about social impacts, and uncertainty regarding future prices of outputs and costs of inputs. As noted above, there is uncertainty regarding the precise discount rate to use. The point of sensitivity analysis is to determine how sensitive the BCR or NPV is to varying discount rates, prices or social or biological effects. If the same policy alternative always ranks number one regardless of the specific discount rate or assumption about future prices, then that policy is probably the best. If the NPV is always greater than zero or the BCR always greater than one, regardless of which of several plausible discount rates are used, then the policy is probably economically efficient. Policy analysts often call such an invariance of ranking or feasibility a robust result. This robustness will aid in defending this policy option as the most economically efficient. Such robustness should be communicated to the decision-maker so that she can have more faith in the BCA.

Alternatively, sometimes the ranking of alternatives or the feasibility of a policy does change if significant differences in future prices or discount rates are used. This too is very important for a decision-maker to know, for at least two reasons. First, this suggests there may be very little real economic efficiency difference between alternatives. Second, sometimes the uncertainty regarding a price, cost or impact variable can be reduced by more sophisticated analysis or collection of more data. One purpose of the sensitivity analysis is to isolate the key variables that have a substantial influence on the outcome from those that have only a minor effect. Those that have a major effect would be the ones toward which the analyst should direct more attention, by increasing the sample size or improving the analysis. Thus performing a sensitivity analysis to address uncertainty about key variables can be an important part of an overall benefit-cost analysis.

Chapter 9

Valuation of Environmental Resources and Quality

Benefit-cost analysis requires valuation of both benefits and costs in the same units. If most environmental policy analyses involved just market goods, this would be not be too challenging of a task. However, most important environmental policy issues involve non-market benefits and market costs: that is, the costs are often fairly obvious and more easily quantified, but the benefits of improved air or water quality or increased protection of biodiversity are far more difficult to estimate. The goal of this chapter is to show the well developed theory and empirical methods in economics to estimate the value people receive from improvements in air and water quality, protecting wildlife, fisheries and recreation. Even what are sometimes regarded as market goods, such as water quantity, have non-market dimensions. For example, while urban water supply is priced, water for in-stream flow (and thus in-stream habitat and recreational use) is not. Comparisons of the benefits of urban water supply versus in-stream use will be one-sided unless we estimate the non-market value of the in-stream flow to rafters, anglers and other river users.

At this point we wish to emphasize that non-market valuation uses the same concept of value as is used for market goods. Specifically, the value of a gain for natural resources, whether marketed or non-marketed, is users' net willingness to pay (as described in Chapter 7). The challenge for non-market goods is to determine this in absence of a market price. But do not let the absence of market price mislead you into thinking there is an absence of value. People do have values for air quality or water quality. While they cannot explicitly and directly reveal them in a shopping center, they may reveal them indirectly in their market purchases or other behaviors. For instance, people reveal the value of air quality in an urban area by how much more they pay for a house that is in an area with better air quality. Sometimes people reveal the value of in-stream flow by how much farther they are willing to travel to a river with higher flows. Contributions to environmental organizations and political activism are other reflections of value. In other cases, the policy analyst might create a hypothetical market through the use of a survey. Here the respondent uses this hypothetical market to reveal what they would pay for improvements in water quality if a market was available. Hopefully by the time you have finished this chapter, you will avoid ever thinking that somehow "economists assign values" to the environment. Economists, with the help of other social scientists and statisticians, merely record and organize information on people's behavior to quantify in dollars the value people already possess. Economists no more create value for the environment than did scientists create sub-atomic particles when they first peered through an electron microscope, or astronomers create distant galaxies when they first peered through more powerful telescopes. Much like these sub-atomic particles or distant galaxies, the values have always been present; it has only been recently that we have been able to accurately measure these values.

1. COMMON METHODOLOGICAL APPROACHES TO VALUATION OF QUASI-MARKETED NATURAL RESOURCES

Many natural resources are not sold in competitive markets. For example, irrigation water is not often sold in a competitive market, because federal, state, or local laws restrict its use or distribution, but yet we can derive a value of irrigation water by looking at market transactions surrounding the use of irrigation water. Sometimes trying to observe the economic value of hydropower is also difficult, but since the electricity is eventually sold, we can often infer this value. To estimate the economic value of these quasi-marketed goods, we can use one of four approaches: (1) market or transactions evidence; (2) demand estimation; (3) residual valuation; or (4) change in net income (Davis and Johnson, 1987; Gray and Young, 1984).

1.1 Market or transactions evidence

Market or transactions evidence requires observation of similar sales of resources in competitive markets to infer a value for the resource of interest. For example, to value wetlands the city may own, the analyst might look at what private organizations such as the Nature Conservancy have paid to purchase wetlands. The key requirement for accurate valuation with this approach is finding market sales of resources with comparable characteristics to the publicly owned resource.

1.2 Demand curve approach

A demand curve approach can be employed if there is a large enough number of past sales, or a data series of prices and quantities sold over time or across the country, to statistically estimate a demand function. Regional or time variation is important in order to get variation in price; without price variation, it is impossible to determine how the quantity responds to price. This approach has been performed for regional and national timber demand (Adams and Haynes, 1980) and for municipal demand for water (Young, 1973, Foster and Beattie, 1979; Hewitt and Hanemann, 1995).

1.3 Residual valuation

Residual valuation involves subtracting from the total value of the output produced all of the costs of inputs except the "unpriced" input, such as water or hydropower. The value remaining after all other costs (including a normal profit, return to management, capital, etc.) is then attributed to the unpriced or underpriced public input. Thus the irrigation value of unpriced water flowing from a National Forest is the value of the agricultural crop minus all of the priced inputs, such as land, seed, fertilizer, machinery, labor and management. The key assumptions here are that all other inputs are priced and that total value of output can be apportioned according to the marginal product of the inputs (Gray and Young, 1984:171).

1.4 Change in net income

Change in net income involves simulation of a firm's or producer's net income with and without additional quantities of the environmental quality. For example, the value of additional water to a farmer is frequently calculated by simulating how the farm enterprise would modify its operation if it had an additional 1000 acre feet of water at a given time of the year. Often this simulation is performed using farm budget data and linear programming models.

1.5 Water as a Quasi-Market Resource

Valuation of water begins our transition from marketed natural resources to nonmarketed natural resources. What is the value of an additional acre-foot of water? The first challenge in valuing this water is determining to which of its many uses this incremental flow will be allocated. Will the additional water be used in irrigated agriculture or municipal uses, or rather be left in the river to add to instream flow? In general, additions to existing water supplies are usually assumed to go to the lowest valued use of water, such as irrigated pasture or irrigated agriculture. The reason is that higher valued municipal and industrial users can normally purchase water from these lower value uses and therefore are already supplied. This is not always the case, however, as rigidities in the prior appropriation doctrine of water law in the Western U.S. gives historic senior rights to the first users rather than to the highest valued users.

After determining which water uses are most likely, the next step is to determine which of the four techniques to use. This is frequently done based on the type of water use to be valued. If the water use is irrigated agriculture, a change in net income or residual valuation approach is often used. Here, farm budgets are used to simulate the increase in crop values stemming from additional water. From the change in crop values the cost of putting this additional water to use and associated farm inputs such as additional seed, fertilizer, tractor time, etc. are subtracted to yield the change in net income.

If the use of water to be valued is municipal, then demand estimation or market transactions will frequently be employed to calculate a value of water in this use. For a discussion of valuing water, see Gibbons, 1986, or Fredricks et al., 1996.

Of course, if the additional water is to be left in the stream to enhance fisheries or recreational boating, then in most areas of the U. S. you must rely on slightly different techniques for valuation of water in these cases. We say "slightly" different because the first technique discussed basically falls under the category of demand estimation. This technique is called the travel cost method and will be discussed in the next section on valuing recreation.

2. VALUATION OF NONMARKETED NATURAL RESOURCES AND ENVIRONMENTAL QUALITY

2.1 Why Nonmarket Valuation?

As we have seen throughout earlier chapters of this book, there is a big difference between financial cash flow and economic value to society. This distinction is even more crucial for public goods and for many uses of natural resources that are not traded in markets. While the economic forces underlying nonmarketed natural resources (such as sport fish, wildlife, and wetlands areas) are similar to those economic forces for any natural resource or good, there are some important differences. Public goods are defined as being nonrival and nonexclusive. "Nonrival" means that use by one does not affect use by another – for instance, my looking at a work of art does not affect your ability to look at the work of art. (Where rivalry exists, my consumption of the good affects your consumption, such as my eating your sandwich.) "Nonexclusivity" means that nobody can be denied the benefits of the good. For instance, many people value preservation of biodiversity in tropical rainforests, even if they never intend to visit the forests or use their products. (Exclusivity exists when I, for instance, can limit your access to the air I have kept clean.)

The nonrival and nonexclusive nature of consumption of pure public goods means that the producer cannot exclude nonpayers from consuming public goods. As a result, the optimum supply is typically financially unprofitable in private markets, even if it is economically viable to society, because a private producer cannot arrange payment from everyone who benefits from the action. Unfortunately, without an explicit or complete market, direct observation of willingness to pay (i.e., benefits) through individual behavior in markets is made difficult. Thus there is a need for techniques to reveal the demand and value for publicly provided natural resources.

The other need for valuation of natural resources is that, institutionally, most Western States have taken ownership of sport fish and wildlife out of the free market. Most game animals are owned by the State, and they cannot be legally bought and sold like livestock. In addition, the State does not ration access to wildlife based on market-determined prices. Even though society has chosen not to allocate access to these resources through the market, economic forces are still present. That is, there is still scarcity of these natural resources, excess demand at these zero or below-cost prices, and competing marketed uses of these resources (e.g., forage that is demanded by ranchers for livestock instead of wildlife, or timber for wood products rather than wilderness recreation). Given the scarcity, excess demand, and competing demands, allocation decisions about uses of these nonmarketed natural resources are still required. If society is interested in receiving the most benefit possible from these nonmarketed resources, some attention to benefits and costs is necessary.

2.2 Revealed versus Stated Preference Methods

A variety of benefits can be associated with nonmarketed goods, such as clean air, wilderness, or endangered species protection. Some of these benefits involve use of the resource, such as recreational visits to an area, wildlife viewing, or buying property in a scenic place. The value associated with these uses can sometimes be calculated from observations on how people behave – for instance, how often they travel to a distant place, or how much they are willing to pay for a particularly attractive property. People "reveal" their preferences for these nonmarketed goods through these activities; thus, so-called "revealed preference" methods of valuing nonmarketed goods use actual behavior to identify values.

In contrast, some nonmarketed goods are difficult to associate with any use value. For instance, many people who have no interest in traveling to the Arctic or to tropical rainforests nevertheless seek to protect wild areas in both places. Though some of the motives for preservation (such as "option value," preserving the option of making use of the area in the future, for visitation or for possible medicinal drugs from plants or animals) involve possible future uses, many of the motives involve no explicit behavioral trace. "Existence value," the value people get just from knowing that wild places or exotic species exist, or "bequest value," the desire to leave these goods for future generations, do not typically have a use or an action associated with them. Nevertheless, these values are real, as reflected by the efforts of large numbers of people to protect these areas. Because people cannot always reveal their values directly or indirectly through their market transactions, economists use "stated preference" methods – methods that rely on how much people say they value the resource, rather than rely solely on their transactions – to measure these values. Stated preference methods are also used to value use values of non-market resources such as recreation or improvements in environmental quality. Stated preference methods are sometimes more flexible than revealed preference methods to provide *ex ante* levels of environmental quality proposed in some policy but not yet experienced.

2.3 Revealed Preference Methods

Revealed preference methods rely on inferences about the value of environmental quality based on actual behavior of individuals in market purchases that are related to environmental quality. The two most common applications are the hedonic property method and the travel cost method. In the hedonic property method, differences in housing prices are used to infer what higher air or water quality is worth to households. With the travel cost method, the cost of visiting a recreation site with, e.g., higher water quality allows estimation of how the demand curve shifts with changes in water quality. We now discuss each of these methods in detail. While this discussion may not give you all the information you need to make you a practitioner, we hope that it will enhance your understanding of these techniques and help you to interpret studies using these methods. Freeman (1993) gives more detail and references on all these approaches, for those who seek to become experts.

2.3.1 Hedonic Methods

2.3.1.1 Hedonic Property Method

The hedonic property method is based on the idea that the value of a house can be decomposed into the value of its individual characteristics – where it is located in relation to workplaces and amenities, how many bedrooms and baths it has, how large it is, and the nature of its surroundings, including environmental attributes. It has been extensively used to estimate the value households living in urban areas place on improving air quality and water quality. One of the most famous studies is the value of reducing sulfur dioxide and total suspended particulates in the Los Angeles area. This study by Brookshire, et al. (1982) compared house prices in areas of Los Angeles that had markedly different levels of air pollution. They found that a house in the best air quality (lower levels of total suspended particulates and nitric oxides) was worth significantly more than a house in an area with poor air quality, even adjusting for the other attributes of the house structure and neighborhood. Converting the house price differentials into monthly rental equivalents yielded estimates ranging from at least $48 per month to as much as $173 per month for an improvement from poor to good air quality (figures are in 1980 dollars; see Brookshire, et al., 1982:173). The same approach has been used for water quality, comparing house prices on shorelines with poor water quality to those with good water quality (d'Arge and Shogren, 1989; Leggett and Bockstael, 2000).

The basic form of a hedonic property price function is:

House Price = f(structural characteristics of the house, neighborhood characteristics such as schools and crime rate, environmental characteristics such as air quality)

For example, we might have the following:

(9-1) House Price = Bo+B1(House Size) +B2(House Age) +B3(Miles to Work) +B4 Air Quality

By collecting data on houses from several areas within a city in which the air quality varies, a separate coefficient on Air Quality (AQ) can be estimated using multiple regression. This coefficient gives the present value of the household's willingness to pay for a one unit change in air quality. The value of a given increment can be approximated by calculating the change in house price with the change in air quality resulting from the program. That is, calculate the house price by inserting the new level of air quality, then again at the without-program level of air quality, and then take the difference in house price. We say approximate, as this estimate of the change in value is accurate only for small changes in air quality. As shown by d'Arge and Shogren, for large gains in environmental quality the regression coefficient will overstate the value of gains. Conversely, the regression coefficient will understate the value of losses.

To more accurately value larger or non-marginal changes in environmental quality, such as air quality, with the hedonic property model, the analyst can estimate a second stage or separate demand for environmental quality relationship from the first stage shown above. The second stage or demand for environmental attribute is:

(9-2) $Qty_{(AQ)}$= Ao-A1$Price_{AQ}$+A2$Price_{subs}$-A3$Price_{comp}$+A4Income

where the prices are implicit prices calculated from the first stage hedonic price function, and AQ, subs and comp are the environmental quality attributes, substitute and complementary attributes for environmental quality in housing decisions. Substitutes for environmental quality might be square footage of housing space, while a complement in a lake water quality study might be lake frontage of the property (Boyle, et al., 1999). Once this function is estimated, the net WTP for non-marginal changes in environmental quality can be calculated as the area under the demand for the characteristic between the original and new level of environmental quality.

The estimation of the second stage or characteristic demand function is more challenging in terms of assembling the data and in the statistical analysis. The data challenges involve the need to collect housing price data in multiple markets in order to econometrically identify the demand for the housing characteristics and deal with possible endogeneity of implicit prices of housing attributes. Specifically, the functional form of the first stage hedonic price function will determine the form of the implicit prices to be used in the second stage characteristic demand function. If the first stage hedonic price function is estimated in the linear form shown above, then the marginal implicit price does not depend on the level of the quantity of the environmental quality attributes. As such the implicit price is constant and does not vary within a single housing market. Thus, to obtain variation in the implicit prices of the environmental quality attribute one must have multiple housing markets which will each have their own prices. Alternatively, if the first stage hedonic price function is non-linear, the implicit price will vary depending on the quantity of the environmental attribute, but this makes the price endogenous. This too creates problems when attempting to estimate the demand for environmental quality attribute, as classical regression techniques require the independent variables to be exogenous. Further, the implicit prices are simultaneously determined by the supply and demand for the characteristic, so the econometric problem of identification has to be solved. Again, the multiple market approach can contribute to solving this problem as does the use of an instrumental variable approach to estimating the demand equation. In the instrumental variable approach, the analyst use a separate equation to forecast the value of the price variable, conditioned on a set of independent variables. This forecasted level of the price variable is then used as the price variable in the second stage hedonic demand function. The use of the forecasted value of the price variable eliminates the endogeneity of the price variable in the hedonic demand function. For more information on the hedonic method see Freeman (1993) or Palmquist (1991).

A variant of the standard hedonic property method described above is to use the repeat sales or resales method. If an important discrete environmental change has occurred in a geographic area, it may be possible to uncover the economic value of that change by comparing house prices before and after the environmental change, controlling for other factors unrelated to the environmental change. For example, the economic value to a family from avoiding a flood might be measured by comparing the sale price of a house after a recent flood in the area to what that same house sold for prior to the flood event. This requires a large enough real estate market that one can observe the same house selling at the two different points in time. Besides flooding, this technique can be used with earthquake hazards, discovery of new hazardous waste sites, fire risk, etc. Essentially, the analyst needs to find a natural experiement, where the risk was not apparent to home buyers before the event as compared to after the event. The economic value of avoiding this new environmental degradation will be capitalized into the housing prices in the repeat sale.

In either case, the aggregate value of the improvement in environmental quality would be calculated by multiplying the change in house price by the number of residences affected by the change. If the value of environmental quality changes with the household's distance from a particular point source of pollution or natural resource, then this needs to be explicitly accounted for in the estimation of the first stage by including a distance variable (e.g., Leggett and Bockstael, 2000). Then a spatial valuation gradient would be used to more accurately arrive at the total value to residences in the area.

The hedonic property method is quite powerful and can be quite convincing to decision makers, as it is based on actual behavior of affected households living in the area. However, besides the traditional concerns of any method relying heavily on statistical estimation (e.g., is the right functional form used, are there any omitted variables that are correlated with environmental quality; see Chapter 6), there is the potential for underestimation due to measuring only residential household values. Not everyone who benefits from an improvement in urban air quality lives in the city, for instance. Some simplifying assumption would be needed to determine how to generalize this value to renters in apartment buildings and non-resident visitors to an area. That is, improvements in water quality at a lake in town is likely to have benefits to more than just the residents living around the lake, but also to those who might bike or drive to the lake for fishing or a picnic lunch.

2.3.1.2 Hedonic Wage Method

Another form of hedonic analysis -- the hedonic wage approach – should also be mentioned here. The underlying idea of hedonic wage analysis is similar to hedonic property analysis: here, wage is decomposed into the attributes of the job, including environmentally-related attributes. In particular, the analyst observes how much additional wages must be paid to an employee to have the employee accept higher risk of illness from air pollution, noise pollution, or other occupational dangers in the workplace. This method is particularly suited for valuation of workplace hazards, a common occupational/environmental policy question. The estimates of how much additional workers must be paid to accept higher illness or death rates is also used to calculate a value of statistical life. For instance, if an additional $1000 per year in salary is required to induce workers to accept a 0.0001 increase in death (one in ten thousand) per year, the value of a statistical life is found by dividing the change in annual income for a higher risk by the change in the risk: in this case, \$1000/0.0001 = \$10,000,000 as the value of a statistical life. This value is often used, not only for workplace safety, but also for valuing roadway safety, health risks associated with pesticides on food, airline safety regulation, and other programs affecting the probability of someone dying. Using a workplace value of a human life for other programs is not a perfect match: workplace hazards are voluntarily accepted by the worker for higher wages, whereas the increased risk of death from lung cancer due to breathing air in the city where one lives is an involuntary risk. Nonetheless, this is a commonly used tool for valuing regulations that save human lives. As with any method, there are numerous simplifying assumptions that are required for wage differentials to accurately reflect what workers are willing to accept for increased risk of injury or death. One of these is perfect information about the relative riskiness of alternative jobs, so that a fully informed trade-off of risk for money can be made. However, in some cases dealing with chemicals or substances with long latency periods (e.g., asbestos), the job risks are not known to the worker (or even to the industry for that matter), at the time the wage-risk trade-off is made. Thus the method is better suited for chemicals or risks which are rather obvious or are well known to workers. For more detail on the theory and mechanics of this method see Palmquist, 1991, Freeman, 1993 or Viscusi, 1993.

The hedonic techniques are very useful and accepted methods of estimating the benefits of many facets of environmental quality. The hedonic property method can be used to value changes in air quality, water quality, stream restoration, wetlands, parks, beaches, open space, noise, and even access to public transportation. If, however, the change in environmental quality being evaluated occurs in a public land or rural setting, the hedonic property method might not accurately reflect the full user benefits. To measure these benefits, we turn to a demand estimating technique called the Travel Cost Method.

2.3.2 Travel Cost Method For Estimating Recreation Demand And Benefits

The second type of revealed preference method is limited to valuing recreational uses of natural and environmental resources.

Unlike marketed goods that are shipped to consumers, in the case of recreation, consumers must ship themselves to the recreational resource. Therefore, people face substantially different implicit prices (both the cost and the time required for travel) for a recreation site depending on where they live. Typically, people closer to a site (with lower travel costs) will visit a site more frequently than those who live further away. As a result, we can statistically exploit these differences in prices, and the resulting differences in numbers of trips, to trace out a demand curve for the recreation site. Once we have the demand curve, we can calculate the net willingness to pay or consumer surplus from it, as described in Chapter 7.

2.3.2.1 Calculating a Demand Curve Using the Travel Cost Method

To calculate a demand curve for recreation at a specific site, we first recognize that travel costs are prices. Figure 9-1 provides a schematic map of 4 cities where people live who visit a particular state forest campground shown in the center of the map. The map shows the trip costs (transportation costs and travel time costs) to visit the campground from four cities of equal population size. People living in each city face a different "price" to visit based on the distance from the state forest campground. As a result, we find people in the close city (#1) take more trips than those from the more distant city (#4). This is as one would expect, holding other factors such as income and substitutes constant across cities.

Figure 9-2 shows how the dollars per trip (shown on the vertical axis) and the number of trips per city (on the horizontal axis) from each of the four cities allow one to trace out a demand curve for this state forest. That is, each trip cost and number of trips per city is a price-quantity combination. If we have adjusted for other socioeconomic differences, such as income (by including them as demand shifters) and availability of substitute campgrounds available to each city, these price-quantity points trace out a single demand curve for camping at this state forest.

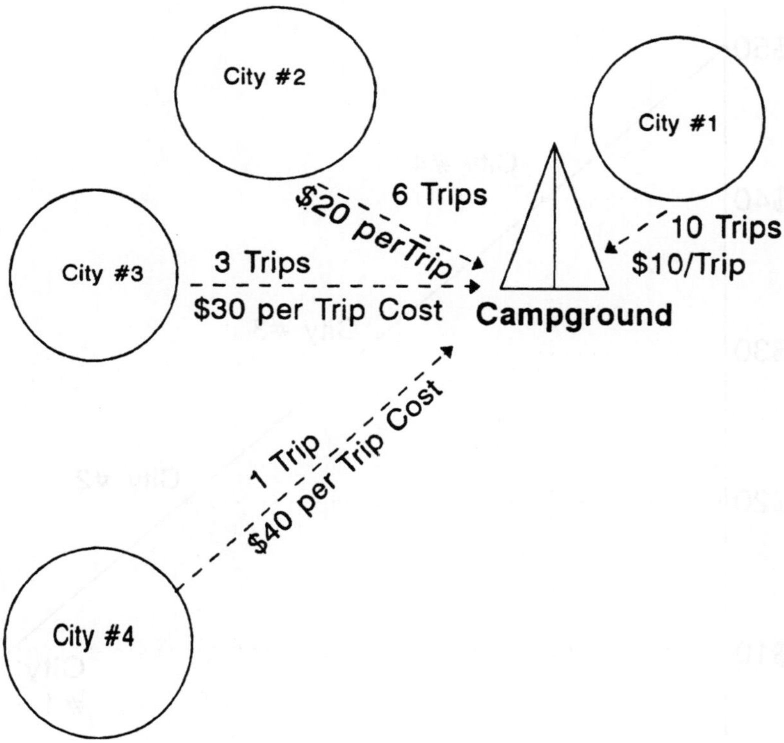

Figure 9-1 Spatial Recreation Market for Camping

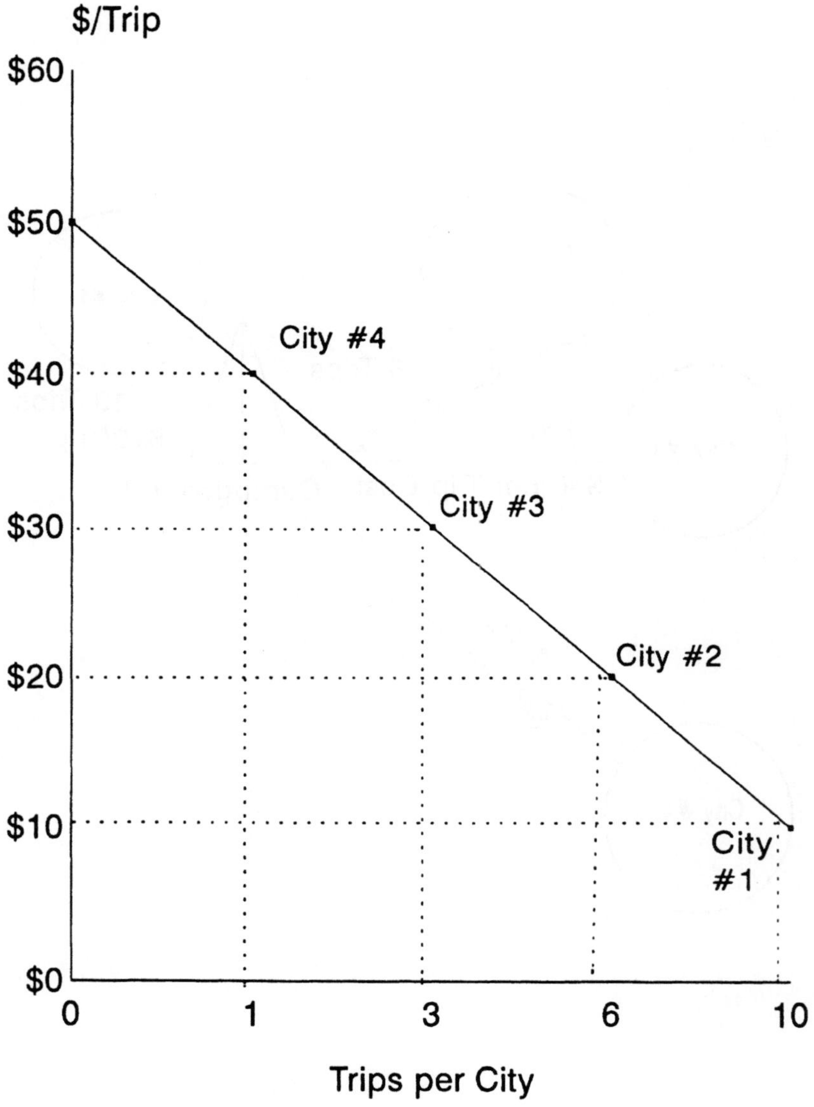

Figure 9-2 Travel Cost Method Demand Curve for Recreation

This demand curve could be estimated with the following basic form:

(9-3) $Trips_i / Population_i = B_o - B_1(\$Travel\ Cost_i)$

where

Trips is the number of trips from city i

Population is the human population of city i

Travel Cost are the expenses involved in travelling to the site

B_o is the horizontal intercept of the demand curve

B_1 is the slope of the demand curve, i.e., the rate of change in trips per capita with a $1 change in travel cost

This exact value of the B_o and B_1 coefficients would be calculated using a least squares regression technique described in Chapter 6.

2.3.2.1.1 From Demand Curve to Benefits

As described in Chapter 7, consumer surplus is the correct measure for the benefits associated with a particular area. Figure 9-3 illustrates how we use this one demand curve to calculate the consumer surplus that campers from each city receive when they visit this site. Figure 9-3 repeats the same demand curve four times, once for each city. This makes sense, since they all visit the same state forest campground and consume essentially the same recreational experience. Since campers located in each of the four cities face a different trip cost or price, the consumer surplus area is different for each city. The consumer surplus triangle is the area under their demand curve but above their current trip cost or price. Those living close to the site buy more camping trips and pay less per trip. Hence they receive a much larger consumer surplus. In other words, they would be willing to pay more to have access to keep the campground open or to prevent the state forest from being logged compared to those living further away, other things remaining equal.

Figure 9-3
Calculating Consumer Surplus From TCM Demand Curve

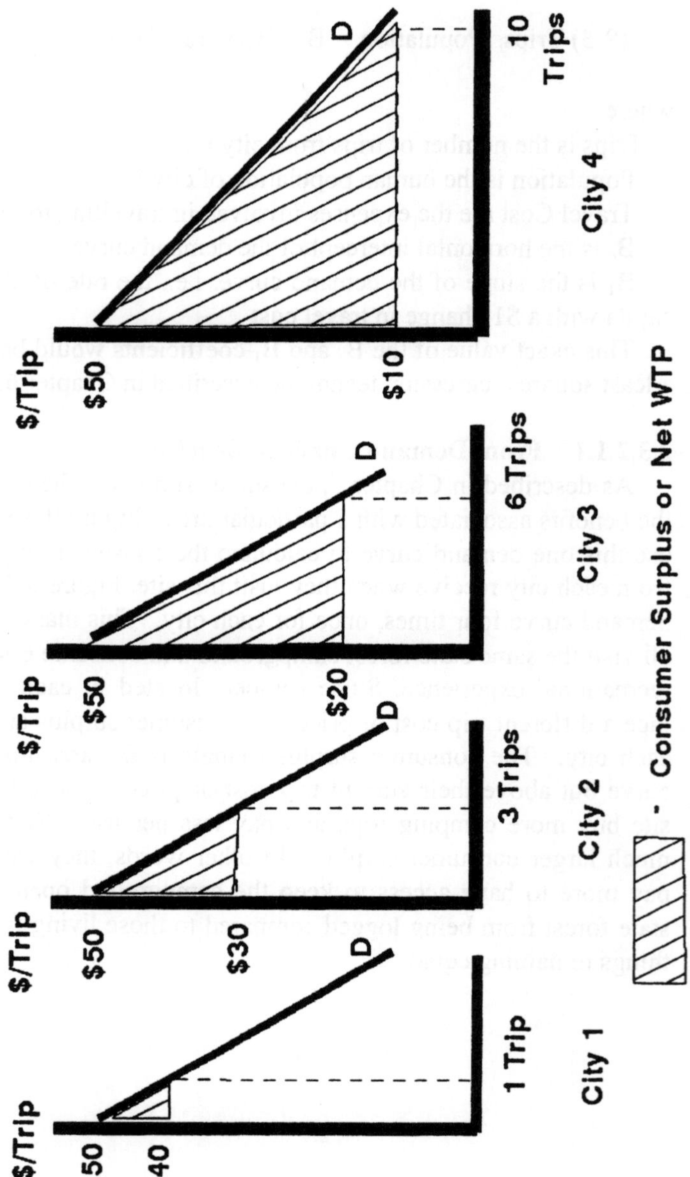

2.3.2.1.2 Calculation of Total Site Consumer Surplus

The total camping benefits are found by adding up all of the trips from the four cities (e.g., 20 trips) and adding up the shaded consumer surplus areas under each city's curve. This total consumer surplus, or net willingness to pay for this site, is shown in Figure 9-4. The vertical axis on this figure is the added dollars of willingness to pay or consumer surplus over and above the visitors' actual expenditure. Thus, trip costs have already been netted out. The horizontal axis is the sum of all trips taken by the four cities. The total consumer surplus received during each recreation season is $325.

The average consumer surplus per trip is $16.25 ($325/20). The $16.25 is sometimes referred to as "average demand price," as it is the average net willingness to pay for a trip. This means a typical visitor would receive a personal profit of $16.25 from visiting this site as compared to the next best substitute. In principle this is the amount the camper would bid per trip in a sealed bid, similar in concept to what a logging firm would bid for the right to harvest trees in the state forest. The economic value per recreation visitor day (RVD) that the U.S. Forest Service developed in their 1985 and 1990 Resource Planning Act Environmental Impact Statement for use in forest planning was originally developed from average consumer surplus estimates. In principle, this is what the famous Unit Day Value table developed by the U.S. Water Resources Council for the U.S. Army Corps of Engineers and U.S. Bureau of Reclamation is intended to represent. These two sources of valuation information will be discussed at the end of the chapter.

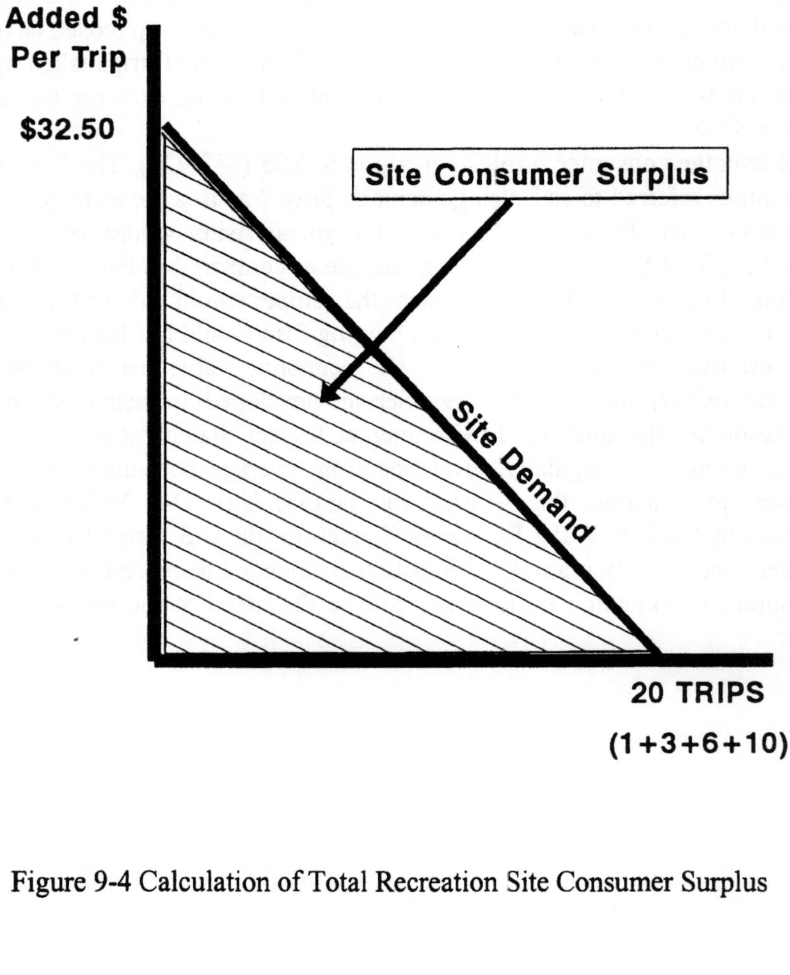

Figure 9-4 Calculation of Total Recreation Site Consumer Surplus

2.3.2.2 Basic Assumptions of the Simple Travel Cost Method

Like any measurement method, whether in economics or biology, the Travel Cost Method (TCM) of estimating recreation demand involves several simplifying assumptions.

2.3.2.2.1 Single Destination Trips

This is one of the most important assumptions. This assumption says that TCM should only be applied to trips taken by visitors for the sole purpose of recreating at this site, and this site was the single destination of the trip from home. In terms of our simple map in Figure 9-1, this means that visitors go from their city of residence to the state forest and return directly home. If this occurs, then it is accurate to measure the round trip distance from home to the site as a measure of the price of visiting the site. If they visit other recreation sites on that same trip, then measurement of the price of visiting any one of the sites is nearly an impossible joint cost allocation problem. Therefore, these visitors are often excluded from the estimation of the demand curve (but the average consumer surplus is applied to them at the end).

There are a few solutions to this problem, including grouping like nearby destinations together as a "mega-site" (Mendelsohn et al., 1992). Parsons and Wilson (1997) allow for differences in the demand of single destination and multi-destination trips by use of a demand shifter and travel cost slope coefficient interaction term. If multi-destination visitors are pervasive, then the contingent valuation method, described below, can be used to value such multi-destination trips. Thus the extent of visitors on multiple destination trips must be researched before the TCM can be applied. In general for many rural campgrounds, reservoirs, rivers, hunting trips, fishing trips, hiking areas, National Recreation Areas, and even state beaches, a large majority (80% or more) of the visitors do take single destination trips.

2.3.2.2.2 Variation in Travel Costs

The second major assumption is that there is sufficient variation in travel costs to trace out a demand curve. In terms of Figure 9-1, we needed campers visiting from cities located a different distances to trace out a demand curve. If all of the campers came from cities located exactly the same distance, all we would obtain is an estimate of one point on the demand curve. With one point there is no way to statistically (or graphically, for that matter) trace out a demand equation. Once again, this is usually not a problem for rural recreation areas. They are located close to a few small towns and at varying distances from many cities and counties.

2.3.2.2.3 No Benefits from Travel

The third assumption is there are no benefits from travel itself. That is, we assume the trip cost is incurred to gain access to the benefits of the recreation site. If people enjoy the travel en route, then part of the travel cost is paid for the en route sightseeing and is not solely a sacrifice made to gain the benefits of the recreation site. If people do enjoy the driving en route, then our estimate of recreation site benefits will be overstated. The empirical evidence on this assumption is mixed. For short drives to infrequently visited sites in scenic areas, such as Colorado or the Pacific Northwest, there does seem to be some sightseeing value during the first hour of the drive. After the first hour or so, the disutility of travel time begins to dominate. In hot and arid climates such as New Mexico, even the first hour of travel seems to have a disutility. Of course, when recreationists make frequent trips up the same roads to visit the same sites, it is reasonable to assume there is probably little sightseeing value. Nonetheless, this is an assumption that, with properly worded questionnaires, can be empirically tested for the recreation site in question.

When the assumptions of the TCM are met, the method is expected to yield valid estimates of visitors' net willingness to pay. When the assumptions of the TCM are not met, the method should either not be applied, or much more complex approaches are needed to address these complications.

2.3.2.2.4 Two Ways the Travel Cost Method Demand Curves Reflect Presence of Substitute Recreation Sites/Opportunities

The actual number of trips taken by campers to this particular state forest in our example reflects the presence of other substitute campgrounds. If those other campgrounds were not present and this was the only campground, more campers would visit this state forest. Thus the horizontal intercept of the demand curve would be further to the right if no other substitute sites existed. The vertical intercept of our demand curve is shifted downward due to the availability of substitute campgrounds as well. This means the maximum amount a camper would pay to visit this particular campground is capped at the trip costs to similar quality (as perceived by the visitor) campground. In the travel cost method, this downward shift in the demand curve is reflected by a substitute index variable. If this index rises when the number of other campground accessible to a given city increases, then the variable will have a negative sign in the demand equation.

The net result is that the consumer surplus calculated from this demand curve reflects the presence of substitutes. In essence the consumer surplus obtained from a TCM demand curve that includes a substitute variable is the net willingness to pay for this preferred site over and above the next best substitute. That is, the benefits are the additional personal profit realized by having access to this site as compared to a less preferred or more expensive site.

These examples reflect one of the key advantages of the TCM: its ability to generate values that are based on actual observed recreation behavior. In addition, another advantage of TCM is that it can often be performed using existing data such as campground fee receipts, recreation permits for hunting, fishing, wilderness, and rafting, as well as simple license plate surveys (to identify how far visitors came to see the site). As such, the TCM is a relatively inexpensive method to calculate site specific recreation values.

2.3.2.2.5 Multi-Site Travel Cost Models to Incorporate Environmental Quality as a Demand Shifter

The single site TCM demand estimating model is quite useful if the analyst simply needs to estimate the value of existing recreation. This is the relevant value if there is a mutually exclusive use of the site that would completely displace the current recreation. For example, if our state forest was clear-cut or leased to a surface mining company, the consumer surplus estimate would show the economic loss to consumers from losing this campground. In the case of river recreation, if the river would be completely inundated by a dam, it would be important to compare the value of lost river recreation with the gains in water supply and reservoir recreation, as well as cost of the dam, to determine if this is an economically efficient change.

However, many policy analyses involve estimating the value of a change in quality of recreation. For example, relicensing of a hydropower dam often focuses on how much additional instream flow should be required to be released during the recreation season. Thus the current recreation will increase in value; the increase in value must be calculated and then compared to the opportunity cost of reduced hydropower. Many proposals to reduce non-point source pollution will improve water quality for fishing and water contact sports. Again, the key issue is measuring the gain in value of existing trips and additional trips that would result from the improvement in water quality.

With a single site TCM demand estimating model, it is difficult to observe how visitation would change with an improvement in water quality, since visitors from all the different cities receive the same water quality at the site. That is, water quality is a public good, which is at the identical level for all visitors to this site. What we need is to observe how visitation changes with changes in water quality.

There are two ways we might observe how visitation changes with environmental quality. If environmental quality varies over time in a predictable fashion at this site, it may be possible to estimate the TCM demand model on a weekly or monthly basis, to reflect how visitation changes with this quality. While this is technically possible, it is rare to have such predictable changes in environmental quality that are known to visitors ahead of time, and to have visitation data on such a small time scale as a week.

The most common approach to incorporating environmental quality into a TCM demand estimating model is to pool visitation data across recreation sites that have markedly different levels of environmental quality. It is quite common for fish catch rates, water quality, instream flow, wildlife populations, and scenic quality to vary from one area to another based on differences in geology, source of water, weather conditions, and management activities. The analyst can take advantage of these natural experiments to observe how visitation responds to these changes in environmental quality. By pooling visitation data across sites (essentially stacking each site's city of origin and site destination data vertically), the following quality augmented TCM demand equation could be estimated:

(9-4) $Trips_{ij}/Population_i = Ao - A_1(TravelCost_{ij}) + A_2(Environmental\ Quality_j)$

where

$Trips_{ij}$ is now Trips from City i to Site j.

Travel $Cost_{ij}$ is now Travel Cost from City i to Site j.

Environmental $Quality_j$ is the level of environmental quality at site j.

The size of A_2 will indicate how much the demand curve shifts to the right or left with a one unit change in environmental quality. For example, if environmental quality is water clarity measured in feet, then A_2 would indicate how much trips per capita would increase with a 1-foot increase in water clarity. The area between the pre-policy and post-policy higher demand curve would be a measure of the economic value of this gain in water clarity. These annual benefits could be compared to the annual costs of increasing water clarity by one additional foot to determine if it is economic efficient action.

2.3.2.2.6 Instream Flow TCM demand example.

The following example illustrates how fishing visitation data were pooled across rivers in Montana to observe how the number of fishing trips changed with fish catch rates. Since fish catch rates are influenced by water quantity and water quality, we essentially have a simple bio-economic model. This model can be applied to estimate the benefits of increased instream flows or improved water quality from reducing non-point source pollution from agriculture or timber harvesting, heavy metal discharge from mine sites, or any other factor influencing fish populations via water quality or quantity changes.

The data used with the Travel Cost Method to estimate willingness to pay for recreational fishing were collected from two surveys of Montana anglers in 1985. The model reflects pooling of visitation data to several rivers in western Montana. The TCM demand equation is:

Table 9-1. Statistical estimates for recreational fishing

Variable	Coefficient	t-Statistic
Intercept	2.471	-2.81
LRTDIST	-2.619	-53.387
LSTROUTC	0.246	3.897
LYRSFISH	-0.885	-8.53
LEDUC	1.492	4.18
LSUBTRTC	-0.017	-3.221
LSOTHRC	0.079	-1.955

where:

LTRIPCAP, the dependent variable = log of trips per capita from county of angler origin i to river j

LRTDIST = log of round trip distance from county i to river j

LSTROUTC = log of sum of trout catch at river j,

LYRSFISH = log of average years fished of anglers in county i

LEDUC = log of average years of education of anglers in county i.

LSBUSTRTC = log of substitute site fishing index. Index is based on trout catch per mile at alternative river k with higher fish catch per mile than the study river j.

LSOTHRC = log of sum of other sport fish catch at river j (mostly whitefish).

The R-squared is quite high, indicating that nearly 82% of the variation in trips per capita is explained by the set of independent variables. The equation also contains statistically significant variables for the influence of substitute rivers and fish catch. The individual coefficients are significant at the 95% level or greater. The coefficient on distance (the price variable) is highly significant.

Consumer surplus estimates for improving water quantity and water quality can be estimated by working with biologists to translate the changes in water parameters into changes in trout populations (this is no easy task, as any biologist will tell you). Once the linkage between water quality or quantity and trout populations is established, the next step is to use the resulting change in the fish catch variable in the TCM demand equations estimated above to calculate the corresponding change in benefits. A river's total consumer surplus with an improvement in water quality is estimated with the site's existing trout catch; then catch is increased by the gain in catchable trout expected under the improved water quality scenario. The gain in the catchable trout variable in the demand equation shifts the Travel Cost Method demand curve outward to D_1 High Fish, as illustrated in Figure 9-5. The area between the D Low Fish and D_1 High Fish can be thought of as the gain in fishing benefits with the water quality improvement.

This shifting demand curve process is repeated for each decade as the fish population increases in response to improved water quality. The present value of the change in fishing benefits is calculated as the present value of the area between the two demand curves in Figure 9-5 over the affected river over some time period, often 50 to 100 years. The present value of these benefits would be compared to the present value of foregone timber or mining, or whatever the policy action is to bring about the change in water quality.

For more details on these multi-site Travel Cost Method models see Ward and Loomis, 1986; or Loomis and Walsh, 1997.

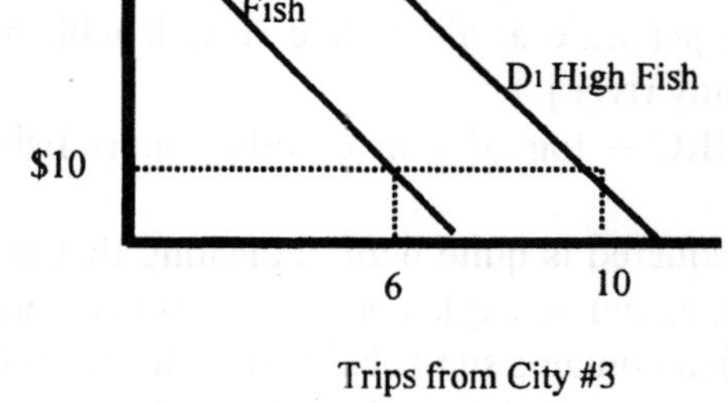

Figure 9-5 How Fishing Quality Shifts Angler Demand Curve

2.3.2.2.7 Recent Variations on Travel Cost Models to Better Incorporate Environmental Quality

In the last several years there have been several variants of TCM developed. These include site choice models, frequently referred to as Random Utility Models. These models are quite similar to the pooled multi-site models described above. The key difference is that the choice problem is facing the visitor is broken up into several stages: first, which site to visit; and next, the number of times to visit. The travel cost and the relative site quality affect both the probability of choosing a given site and the number of times the individual will visit each site. The site choice problem uses a more complex form of multiple regression, such as multinomial logit models common in the field of transportation planning. See Herriges and Kling, 1999 for a discussion of the several types of Random Utility Models.

Finally, there are also hedonic travel cost models that reverse the viewpoint of our original TCM models. In these models, the visitor is viewed as determining how much more travel cost to spend to "purchase" recreation areas with higher quality levels.

All of these models allow for valuation of the incremental change in environmental quality and hence are quite useful in policy analysis. However, these models often require quite detailed survey data and sophisticated statistical analysis. For evaluation of regulations with multi-million dollar costs, such an analysis effort is more than justified. But for small policy analysis, the cost of estimating these more elaborate models can quickly absorb the entire project analysis budget. For small projects we discuss the benefit transfer approach at the end of this chapter.

2.4 STATED PREFERENCE APPROACHES

As discussed above, the revealed preference methods are based on observing decisions involving a market component (cost of property or travel to a recreation site) and using that market information to price non-marketed goods. It has the tremendous advantage that it is based on actual behavior: people are putting their money where their mouths are. Also as mentioned above, though, these methods *only* capture the values associated with those who participate in the associated market. For some environmental goods, such as biodiversity, climate change, or protection of endangered species, there may be no obvious associated market, or many who appreciate provision of the environmental good do not participate in that associated market. For instance, a travel cost study to assess the value of protecting Prince William Sound (the site of the Exxon Valdez oil spill) from oil damages would not capture the value of many people who do not travel to Alaska but nevertheless would like to see the Sound protected. For these so-called non-use or passive use values, the Contingent Valuation Method (CVM) and conjoint analysis (CJ) are the primary tool for valuation. In addition, CVM can measure the use value of a change in environmental quality prior to the policy being implemented. This is a useful feature in a policy analysis, especially of actions that are quite costly or possibly irreversible.

2.4.1 Contingent Valuation Method

The contingent value method (CVM) is a survey technique that constructs a hypothetical market or voter referendum to measure willingness to pay or accept compensation for different levels of nonmarketed natural and environmental resources. The method involves in-person or telephone interviews or a mail questionnaire. CVM is capable of not only measuring the value of outdoor recreation under alternative levels of wildlife/fish abundance, crowding, instream flow, etc., but is the only method currently available to measure non-use or passive use values, such as the benefits the general public receive from the continued existence of unique natural environments or species.

The basic idea of CVM is that a realistic but hypothetical market for "buying" use and/or preservation of a nonmarketed natural resource can be credibly communicated to an individual. Then the individual is told to use the market to express his or her valuation of the resource. Key features of the market include: (1) description of the change in environmental quality being valued (2) means of payment (often called payment vehicle) and (3) The valuation elicitation process, i.e, the particular willingness to pay or willingness to accept question format.

2.4.1.1 Value Elicitation Format

The value elicitation procedure can be one of three types. First, there is an open-ended willingness to pay question format that simply asks the respondent to state his or her maximum willingness to pay, for instance, for protection of an endangered species or improvement in water quality. This was one of the early forms of CVM and is still the simplest. The answer to such open-ended questions is the respondent's willingness to pay in dollars.

Another procedure is the close-ended "iterative bidding" type question. Here, the interviewer states an initial dollar amount, and the respondents answer yes, they would pay, or no, they will not pay. If a yes is elicited, the amount is raised and this process repeated until a "no" is recorded. The highest dollar amount receiving a "yes" answer is recorded as the person's maximum willingness to pay. If a "no" is recorded on the first dollar amount, the amount is lowered until the first "yes" is achieved. In this iterative approach, poor selection of the starting value may influence the respondent's final reported value. This starting point problem can often be minimized by selecting the initial bid amounts around the mean of WTP from an open-ended pre-test, or at least avoiding extremely high or low starting points.

For many economists, the preferred value elicitation approach is the dichotomous choice or "referendum." Here, the respondent just answers yes or no to one randomly assigned dollar amount chosen by the interviewer. The dollar amounts vary across the respondents. In this way, the entire sample is asked to pay a variety of dollar amounts. Then the analyst can calculate the probability of a respondent answering yes or no to a given dollar amount. From this probability, the statistical expected value of willingness to pay can be calculated.

This dichotomous choice approach has several advantages. Its "take it or leave it" format matches the market where individuals either buy or do not buy at a given price. In addition, when cast as a referendum, it matches how people vote on many public programs. That is, they vote "yes" or "no" at a specified tax price, just as people often do on ballot measures. The dichotomous choice approach avoids the starting point problem associated with iterative bidding. Unlike the open-ended WTP questions, where some respondents have difficulty in providing an exact dollar amount of their WTP, in the dichotomous choice format the respondents must simply determine if their value is greater or less than the dollar amount they are asked to pay. Then, much like the TCM, the analyst uses statistical analyses to infer how much more their WTP is than the amount they are asked to pay. For more details on these three approaches see Cummings, et al., 1986 and Mitchell and Carson, 1989.

2.4.1.2 Means of Payment

There are several ways in which a respondent can be asked to pay his or her WTP. The particular means should be tailored to match the characteristics of the hypothetical market the analyst has set up. For preservation of wetlands, rivers or nongame wildlife habitat, a payment into a trust fund similar to "non-game check-offs" on many State income tax forms is often used. Other payment vehicles include utility bills, higher taxes, or higher product prices. In selecting the payment vehicle, the analyst should pre-test different payment vehicles to see which ones yield the fewest emotional responses that are clear protests to the means of payment.

For whatever payment vehicle ultimately chosen, it is still important to use a check question for people responding with a zero WTP or consistent "no's" in a dichotomous choice format to determine why they are giving that response. Are these people with true zero WTP, due to lack of interest or income, or rather are they protesting some feature of the hypothetical market? For example, people will sometimes respond that some other group should pay, or that government should reallocate more of its existing taxes to provision of this good. Either of those answers indicates a positive value for the resource but a concern over the market portrayed in the survey. These protesting responses are normally omitted from the final analysis of WTP. In contrast, people who really are not willing to pay for the resource because they are not interested in it or do not consider it important can be included in the study, since zero really is their value.

2.4.1.3 Description of Resource to Be Valued

Because contingent valuation attempts to simulate a market or voting setting, having informed consumers or voters is expected to produce better, more robust decisions. Thus, CVM studies inform people about the tradeoffs they are facing. The natural or environmental resource to be valued can be described to the respondent in words, drawings, photographs, charts, maps, and/or other methods. Often some combination of these approaches is used, depending on the nature of the environmental change being communicated. For changes in air quality or water quality, many times photographs are used. For changes in lake levels, river levels, or acreage of wetlands, often simple maps or drawings are used. For changes in hunting or fishing success, a simple narrative will work, because hunters and anglers are often quite familiar with the resource to be valued. The key in any of these cases is to provide a short and neutral description of the resource. The regional or national significance of the good can also be described as part of the survey.

These three components (the elicitation procedure, means of payment, and resource description) are combined into a survey instrument. The question wording needs to be very matter of fact: if the respondent pays the money they get the improvement in environmental quality; if they do not pay, they do not receive the improvement (i.e., they stay at their current level).

2.4.1.4 Example of Using CVM to Measure the Value of Environmental Risk Reduction

To illustrate the application of dichotomous choice or referendum CVM to estimate the benefits of an improvement in environmental quality, we present the following case study based on Loomis and duVair.

2.4.1.4.1 Risk Reduction as the Environmental Quality Improvement

In this example, the commodity or good to be valued is reduction in risk of premature death, through state-financed incentives for a California program of hazardous waste minimization by private industry. The public provision of a statewide program, and the fact that in California the funding of these types of programs is often decided via a popular referendum, made the dichotomous choice referendum elicitation format quite credible. The use of a voter referendum elicitation format (where people vote yes or no) made taxes the most logical or credible payment vehicle. Unfortunately, the payment of higher taxes is not an emotionally neutral subject for many people, and such a payment vehicle may increase the number of protest responses. However, it is the realistic possibility of this particular payment mechanism that motivated selection of this method.

2.4.1.4.2 Calculating WTP from the Dichotomous Choice Question Format

Although responses to dichotomous choice questions do not directly provide the maximum WTP for risk reduction by households, there are statistical inference techniques to estimate maximum WTP from data on the probability of a YES or NO response to specific dollar amounts. The probability a respondent will answer "YES" to the WTP question is assumed to be related to the expected gain in well being obtained from receiving the health risk reduction, over and above the satisfaction lost due to paying higher taxes (Hanemann, 1984).

To be more specific, a state-dependent utility function is posited where UL and UD are the utility when alive and dead, respectively. Further let PD be the baseline probability of premature death. Baseline expected utility (EU) can be defined as:

$$\text{(9-5)}\quad EU = PD[UD(I)] + (1\text{-}PD)[UL(I)]\,,$$

where I is income.

The proposed hazardous waste minimization program reduces the probability of premature death from PD to P'D, but at a proposed cost to the respondent of $X each year. If the reduction in the probability of premature death from PD to P'D yields more expected utility that the loss of $X in income, the person will answer yes to the dichotomous choice question. Specifically, the expected utility difference (EUD) is given by:

$$\text{(9-6)}\quad EUD = \{P'D[UD(I\text{-}\$X)] + (1\text{-}P'D)[UL(I\text{-}\$X)]\} - \{PD[UD(I)] + (1\text{-}PD)[UL(I)]\}$$

If this expected utility difference is linear in its arguments, and if the associated additive random error term is distributed logistically, then the probability a respondent will answer YES to a question asking him or her to pay \$X for a program that would reduce the risk of premature death from PD to P'D would be:

(9-7) Probability of Yes) = P(Y) = $1 - [\,1 + e^{Bo-B1(\$X)}]^{-1}$

Maximum likelihood statistical routines such as logistic regression can be used to estimate a transformation of this equation in the form of:

(9-8) $\text{Log}\ \{P(Y)/[1-P(Y)]\} = Bo - B1(\$X)$

Estimates of the parameters Bo and B1 allow identification of the cumulative distribution function of WTP for the risk reduction program (Hanemann, 1984). The mean of the cumulative distribution function is the mean WTP. Since WTP for an unambiguous improvement in expected utility is nonnegative, the mean is given by Hanemann (1989) as:

(9-9) $WTP = (1/B1) * \ln(1+\exp(Bo))$.

2.4.1.4.3 Communicating the Risk Reduction

It is crucial that households understand the environmental improvement they are being asked to purchase. There has been much innovation in risk communication devices over the short history of valuation of health-related risk. Mitchell and Carson (1985) as well as Gerking, DeHann and Schulze (1989) have used risk ladders. In these ladders, each rung represented progressively higher and higher risks. The ladder arrays different probabilities of death from a variety of sources, with the most hazardous at the top. The survey used a multi-color risk adder to show a wide range of involuntary risks and provided a perspective on the size of risk from exposure to heavy metals, relative to the other involuntary risks. The risk ladder showed the reductions in risk level from three alternative programs as movements down the ladder and was used to elicit WTP responses. Thus the ladder was directly used to provide perspective on how much safety was to be purchased. A black and white copy of the ladder is shown in Figure 9-6.

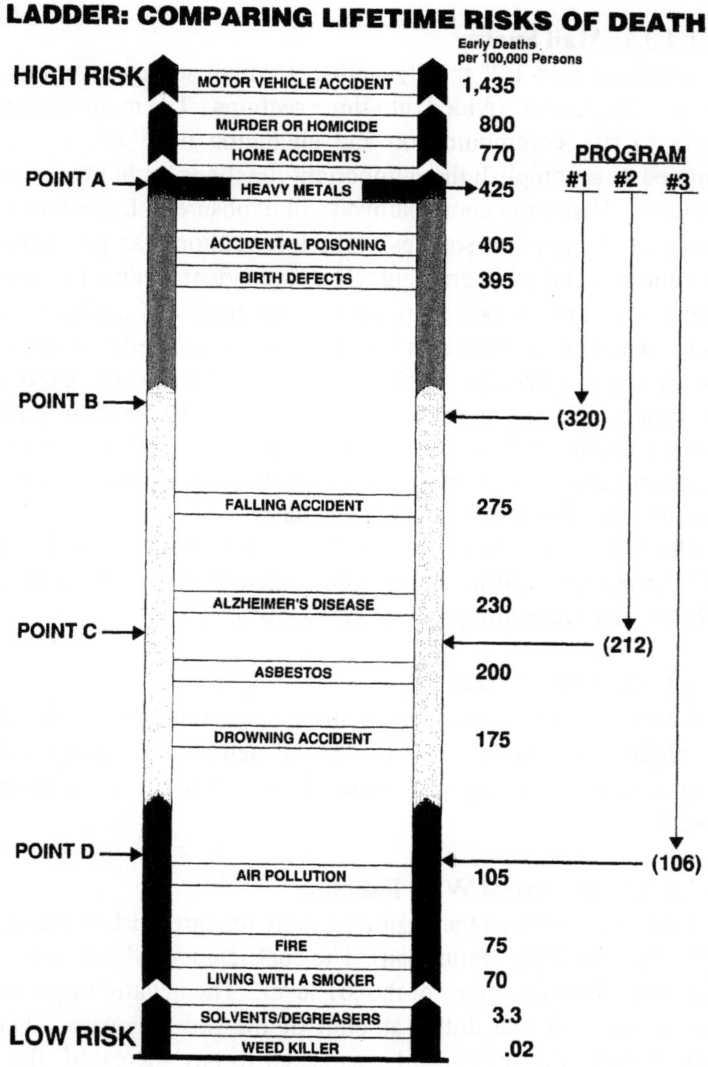

Figure 9-6 Example of Visual Aid Used in CVM Survey on WTP for Risk Reduction

2.4.1.4.3.1 Mail Survey

A full-sized (8-1/2" x 11") multi-color mail survey instrument of twelve pages in length was divided into three sections. The main section of interest relates to risk communication and elicitation of WTP. This section was designed to accomplish three important functions within the survey. First, it provided information about pathways of exposure to hazardous wastes from various contamination sources. The information was presented in written form and in a full-page drawing. The respondent received a description of a current hazardous waste minimization program in California and how the risk of exposure to hazardous wastes could be reduced by greater funding to this program. Besides explaining the pathways of exposure and a mechanism for reducing the risks, the second section contained risk communication devices to convey risk levels. The current risk of premature death from exposure to heavy metals was communicated in narrative and illustrated in the risk ladder.

The final task accomplished in the second section was the elicitation of WTP responses. Respondents were asked three WTP questions, one for each size risk reduction program (25%, 50%, 75%).

2.4.1.4.3.2 Survey Data

A total of 1,000 surveys were mailed to a random sample of California households. A response rate of 43% of deliverable surveys was achieved using a repeat mailing procedure. This yielded a final sample of 374 surveys.

2.4.1.4.3.3 **Statistical WTP Function**

Table 9-2 provides the logit equations for three risk reduction programs (25%, 50%, and 75% reduction). The coefficient on dollar bid is significant in all three logit equations at the .01 level. The negative sign indicates that respondents took the dollar amount of the bid seriously. That is, as the dollar amount the respondent was asked to pay increased, the probability they would pay this amount went down. The other variables also have an intuitive relationship to the probability they will pay for the risk reduction program. This also illustrates the internal validity of the dichotomous choice CVM responses in this example.

2.4.1.4.3.4 Benefit Estimates

WTP is calculated using equation (9-9). The mean of WTP for the three risk reduction programs is as follows: for the 25% reduction the annual WTP is $351 per year, while for the 50% and 75% reductions, the annual

WTP is $449 and $515, respectively. As can be seen, these values are consistent with diminishing marginal value of additional risk reduction.

These values would then be aggregated over the total number of households in California to arrive at statewide WTP. This aggregate WTP would be compared to the annual cost of reducing heavy metal releases to determine which level (25%, 50% or 75%), if any, of reduction was economically efficient.

Table 9- 2. Logistic regression estimates for risk reduction from CVM survey

Variable	25% Risk Reduction		50% Risk Reduction		75% Risk Reduction	
	Coefficient	t-Statistic	Coefficient	t-Statistic	Coefficient	t-Statistic
Constant	-0.7413	-0.757	-0.7907	-0.799	0.27716	0.277
Other Problems	-0.3400	-1.430	-0.2690	-1.135	-0.99747	-3.726
Contacts	0.0505	1.330	0.0577	1.517	0.09043	2.315
Education	0.1042	2.165	0.0929	1.948	0.09440	1.926
$Bid Amount	-0.0034	-5.440	-0.0027	-5.164	-0.00256	-5.556

Where:

Other Problems is a variable reflecting the importance of other problems facing the community relative to the environment.

Contacts is a variable reflecting the sum of individual's likelihood of coming in contact with hazardous materials from consumer product, air, food and water.

Education measures years of education.

$ Bid Amount is the dollar amount of the household was asked to pay each year.

2.4.1.5 Advantages of CVM

There are at least three advantages of CVM. First, the method will work to value recreation even in the face of most visitors taking multiple destination trips. Therefore, CVM can be used to value recreation areas that are visited as part of a longer trip to numerous sites. The second advantage of CVM relates to use of its hypothetical nature as an asset. In many policy analyses, decision makers need information about how people value alternative policies prior to their implementation . For example, what is the value of maintaining a free-flowing river from being dammed or a roadless area from being logged, when these events have not happened? These situations call for information on intended behavior about possible alternatives rather than past behavior if the information is to be timely in assisting the decision maker.

Perhaps one of the strongest advantages of CVM is in valuing those environmental effects for which recreation is but a small part of the social benefits, such as human health effects outside of the workplace or protection of unique natural environments and species that are rarely visited or seen by people. Many people derive a substantial amount of enjoyment or satisfaction knowing these unique areas or species exist, even if they do not visit or see them. For example, many people value knowing that spotted owls, whooping cranes, wolves, condors and grizzly bears exist, even though the may never see them in the wild. This **existence value** is often reflected in a person's contribution to conservation organizations to save these species or natural environments. In addition, some people are willing to pay to preserve areas for future generations. This **bequest value** is an additional motivation. Lastly, some people are willing to pay a premium over their future use value as an insurance premium to insure that natural environments or wildlife species exist so they can visit them in the near future. This is sometimes referred to as an **option value**.

These option, existence and bequest values capture much of what is called the public trust values of natural resources. Others refer to these three values as preservation values (Walsh, et al., 1984) or non-use values (Kopp, 1992) or passive use values (Arrow, et al., 1993). By whatever name they are called, CVM is presently the only method for measuring these values in dollar terms. The U.S. District Court of Appeals indicated these passive use values are appropriate values for measuring losses to the public from natural resource damages arising from hazardous material spills, leaking landfills, old mine sites, etc. The National Oceanic and Atmospheric Administration (NOAA) has taken the same view with regard to losses from oil spills under the Oil Pollution Act of 1990. When a policy analysis involves relatively unique environments or an irreversible decision, these passive use values are often empirically important. For example, in valuing protection of a unique lake in California (Mono Lake), at least 85% of the benefits received by California households were related to existence and bequest values (Loomis, 1988). For more discussion of the use of CVM for valuing existence values see Mitchell and Carson, 1989.

2.4.1.6 Assumptions of CVM

There are two primary assumptions of CVM (Dwyer, et al, 1977). First is that a respondent does have a well-thought out value for the natural or environmental resource of interest. That is, he or she has well-defined preferences for this improvement in environmental quality relative to other goods she buys. The second assumption is that this value can be accurately elicited from the respondent in a survey or interview.

Specifically, the survey can provide the context for the respondents to reveal their true values.

Both of these assumptions are often reasonable for elements of environmental quality such as human health, air pollution, water pollution, noise reduction, food safety, recreation, and well known wildlife species. However, the first assumption may not be fully met for obscure endangered species or protecting remote natural areas. Thus, the more familiar the respondent is with the environmental issue in question, the more likely the assumption is to be met.

For example, visitors to the Grand Canyon, when asked questions about air quality for scenic visibility, will likely have a well-developed value. After all, they have traded money in the form of travel costs (and hence given up other goods) to visit the area. The same is true for many other recreational users. For CVM questions dealing with passive use values for a species or area they may have only read about, the degree of conformity to the first assumption is somewhat lower. Thus the variance around estimates of passive use values is likely to be larger than around estimates of health or recreation benefits.

Nonetheless, much like new product market testing, many economists believe that a respondent can make a judgment about the worth of this resource compared to ones currently being consumed. Especially if the choice is framed in a referendum context, this decision is similar to voting on ballot items to protect some area as a public park.

With regards to the second assumption, much effort in survey design has gone into construction of the hypothetical market that will elicit honest answers. Drawing on past research and pre-testing, a credible hypothetical market or referendum can often be constructed that does provide consequential choices that give respondent the incentive to report their true values. The key is to emphasize that information from the survey will be used by government officials to make decisions about provision of the public good and tax levels. If respondents believe their answers will have real consequences, then truth telling would be the dominant behavior in CVM survey responses (Carson, et al., 1999).

2.4.1.7 Validity and Reliability of CVM

It is reasonable of course, that people might question the accuracy of answers to hypothetical markets as compared to real markets where cash actually changes hands. Would people really pay the dollar amounts they state in these surveys? The empirical evidence to date indicates that, when asked willingness to pay (rather than willingness to accept), people with direct consumption experience of the good being asked about in the survey would pay approximately what they state in the surveys. This conclusion is based on comparisons of actual behavior-based methods (i.e., the revealed preference methods such as travel cost and hedonic methods discussed above) with hypothetical markets used in CVM. One comparison was between housing price differentials in areas of Los Angeles with less air pollution and how much people said they would pay for the same reduction in air pollution. This study by Brookshire et al. showed that stated willingness to pay for a reduction in pollution was statistically less than what people had actually paid.

In a comparison of 80 studies that used both TCM and CVM to value recreation, Carson et al. (1996) found that, on average, CVM resulted in *lower* estimates of WTP than did the TCM. Overall, CVM produced benefit estimates that were about 90% of those produced by TCM for the same recreation resource or recreation activity.

Welsh (1986) used an actual cash market to investigate the validity of CVM. In his study of deer hunting permits, the Wisconsin Department of Natural Resources allowed Welsh to buy and sell up to 75 deer permits. Two markets were set up to allow hunters to buy permits. One was a real market, where hunters without permits were offered actual hunting permits at a specific price and asked if they would buy at that price. The other was a Contingent Value market: the same price was quoted to a different sample of hunters, the transaction did not involve real money, and no permits traded hands. The Contingent Value market obtained values that were generally 25% higher than the actual cash values (Welsh, 1986).

Comparisons of CVM WTP estimates with actual cash in an experimental or laboratory setting have been less encouraging. In several experiments, WTP estimates with open-ended and dichotomous choice CVM were about 2-3 times higher than actual cash (Loomis, et al., 1996). Actual cash validity studies for pure public goods, especially those exhibiting primarily non-use values, sometimes yield even larger differences with actual cash donations (Champ et al., 1997). However, cash donations are not a perfect criterion for judging the validity of CVM WTP estimates, since cash donations are subject to well known opportunities for respondents to free- ride in the actual donations as compared to hypothetical (Carson, et al., 1999). Nonetheless, progress has been made in recalibrating CVM WTP responses using certainty of respondent WTP answers to arrive at valid estimates (Champ, et al.). Other recent innovations that appear to improve the correspondence of CVM WTP estimates to actual cash WTP is a learning design of Bjornstad et al. (1997) that puts the respondent through a prior hypothetical and then a real cash WTP trial prior to asking the actual policy CVM question.

Since validity is difficult to establish unambiguously for passive-use-value-type public goods, reliability of general household CVM responses has been investigated. In a test of the reliability of households' WTP for recreation, option, existence, and bequest values for preservation of a unique wild bird habitat (Mono Lake), Loomis used the test-retest method. Here the same households were resurveyed eight to nine months after their first survey. The surveys were identical, and people asked to give their value in response to both open-ended WTP and dichotomous-choice questions. For both the visitor and general public samples, using both the open-ended and dichotomous choice question format, CVM WTP values proved to be reliable. Specifically, there were no statistical differences between their WTP given in their first survey and the second. In general this indicates the answers are thoughtful responses that are relatively stable over time (Loomis, 1989). This result has been replicated in all published reliability studies that have followed, including the Exxon Valdez oil spill CVM. See Loomis, 1989, and Carson et al., 1997, for details of these reliability studies.

The essence all of these comparisons is that respondents make an attempt to give their value of the resource in a CVM survey. The behavior exhibited and statements of value appear to sometimes understate the value and, in other cases, to overstate the value. Based on these studies, it appears that one can have some confidence that statement of willingness to pay elicited in Contingent Value surveys bears a close resemblance to the behavior that would be observed if the situation described in the survey arose in a real market. Although CVM estimates often do contain error, many economists prefer to use an approximate estimate of these conceptually important public good values than to ignore such values in policy analysis. Failure to quantify the public good values often implicitly results in them have a zero value in the policy analysis. While a well-crafted policy analysis would draw attention to unquantified values, failure to quantify public good values can make them appear less important and certainly more difficult to compare to quantified costs. While CVM estimates of WTP for passive-use values may contain error, the error often pales in comparison with the error of implying a zero value. Nearly all CVM studies published in the peer reviewed scientific literature find that passive use values are statistically greater than zero. While there is clearly room for improvement with CVM, as with any other economic or non-economic model, it has been judged sufficiently reliable by a blue ribbon committee chaired by two Nobel Laureates to be used as a starting point for providing estimates of passive use values (Arrow, et al., 1993).

2.4.2 Conjoint Analysis

Market researchers have used a stated preference technique called conjoint analysis for decades to analyze consumers' trade-offs between market good attributes such as safety, durability, prices, color, style, etc. In the last decade this technique has recently been applied to the estimation of preferences for non-market goods. Conjoint analysis is based on methods for analyzing complex decision-making. These methods are based on a body of research called information integration theory (or IIT) (Louviere, 1988). Conjoint analysis shares with CVM a foundation based on stated preference, what people say they would choose or do in a survey setting. As such, conjoint analysis is capable of measuring both direct use values (e.g., recreation, air quality, health) as well as passive use values (e.g., existence, bequest values). Also like CVM, there are different preference elicitation formats.

We will discuss the basic ratings approach, where a consumer rates each bundle or package of attributes on a scale such as 1-10. Alternative question formats include having survey respondents rank each product or program, or even simply choose the best one. The focus on a product or program as a bundle of attributes suggests that conjoint also has some similarity with hedonic methods. The key difference is that with conjoint, the policy analyst is free to vary the number and quantity of environmental attributes to match the policy question being evaluated. Since the hedonic property method is revealed preference, the analyst can only use the types of attributes and levels of these attributes found in the particular housing market of interest. Often some of these attributes are correlated, making it difficult to estimate the separate effect of each attribute. With a conjoint experiment, the policy analyst can vary the attributes and reduce the multicollinearity so as to allow for more precise estimation of each individual attribute's coefficient and resulting marginal value.

2.4.2.1 An Overview of the Theoretical Foundation of Conjoint Analysis

Conjoint analysis is consistent with the Lancasterian view of consumers receiving utility from goods based on the type and levels of characteristics or attributes available from the goods. The levels of these attributes, and a consumer's knowledge about them, will form the basis for the utility a consumer expects from a particular choice between goods. This utility is unobservable, but one assumption made in conjoint is that it is linearly related to a consumer's responses on a preference rating scale:

(9-10) $U_j = a + bR_j + e_j$

where U_j is the unknown utility of choice j, a and b are regression coefficients, R_j is the rating applied to choice j, and e_j is the random error associated with the regression. The second assumption that must be made is that the consumer uses the rating scale as an approximate interval measure. Finally one must assume that the rating strategy of the consumer reveals his decision strategy.

The rating applied to each choice is based on a consumer's expected overall utility for that choice, which is in turn based on the utility associated with the level of each attribute of the choice. If a choice has three attributes, equation (9-10) can be rewritten as:

$$\text{(9-11)}\quad U_{pqr} = C + u(A_{1p}) + u(A_{2q}) + u(A_{3r})$$

where U_{pqr} is the overall utility associated with choice or good pqr, and $u(A_{1p})$, $u(A_{2q})$ and $u(A_{3r})$ are the "part worth utilities" associated with attributes 1, 2, and 3 at levels p, q, and r, respectively. Equation (9-11) assumes an additive form for the consumer's utility, although other forms can also be used to model the utility for the choice or good.

In a conjoint experiment, each choice presented to the respondent will have different levels of the attributes; these levels are chosen by the researcher. The attributes used in an experiment must be chosen carefully in order to encompass the set of determinant attributes most likely to influence a consumer's preference.

The part worth or attribute values can be estimated by asking an individual to compare two bundles, each with different levels of the attributes. Two attributes in each bundle could (holding all other attributes constant) be varied in opposing directions to reach a point where the two bundles will be equivalent (that is, the individual will be indifferent between them). If one of the attributes is the cost of the good or bundle, then, using the cost of each bundle, an indirect utility function can be derived that allows for monetary valuation. That is, through varying the cost and another attribute to achieve equality between the two bundles, the marginal rate of substitution can be used to represent the marginal willingness to pay for the attribute. In this manner, a value can be estimated for each attribute by varying the level of the attribute and the cost of the option (Mackenzie, 1992). Using regression coefficients, one can calculate a marginal value (or willingness to pay) for each attribute using the negative ratio of the coefficient on the attribute over the coefficient on cost.

Since one cannot observe U_{pqr}, one can use a respondents' rating for choice *pqr* and rewrite equation (9-11) as:

$$(9\text{-}12)\; R_{pqr} = D + u(A_{1p}) + u(A_{2q}) + u(A_{3r}) + e_{pqr}$$

where R_{pqr} is the rating given to choice or bundle pqr, $u(A_{1p})$, $u(A_{2q})$, and $u(A_{3r})$ are as defined above, and e_{pqr} is the random component of the respondent's rating. Ratings are often used to assess consumer preferences in a conjoint experiment, but rankings and binary choice are also used. With ratings as the dependent variable in equation (9-12), the attribute coefficients can be estimated using an ordered probit or ordered logit regression type model.

The above example uses an additive model of utility, that is, it assumes that the attributes contribute to the overall utility of the choice in an additive manner. This means that responses to levels of one attribute are independent of responses to the levels of other attributes. When the number of attributes and/or levels is large, the number of possible combinations (or options) to be evaluated by the consumer becomes unwieldy quickly, especially if interactions are expected to be significant. According to Louviere (1988), the majority of variation in ratings can be explained using main effects or linear additive effects only. Thus it is appropriate to use a main effects experimental design which uses a subset of the possible combinations (Adamowicz, et. al, 1998). Each survey version may include only a fraction of the possible main effects. Thus, if there were 25 possible combinations of attributes and their levels, each survey respondent may only see five combinations chosen at random. Thus the sample would have to be large enough that all combinations were answered by enough people to pool the data and estimate the regression model.

2.4.2.2 Previous Applications of Conjoint in the Environmental Area

Conjoint analysis has been extensively applied to predict consumer (market) behavior. It has recently begun to be used to estimate values for non-market goods. Lareau and Rae (1989) used the technique to estimate willingness to pay to reduce diesel odors. Gan and Luzar (1993) applied conjoint methods to the problem of valuing waterfowl hunting. Revealed preferences and stated preferences (via conjoint analysis) for water based recreation were compared by Adamowicz et al. (1994). Stevens, et al. (1997) examined the value of various groundwater protection programs using a conjoint technique. Finally, Zinkhan et al. (1994) offered an example of the use of conjoint analysis to estimate the multiple values associated with forest management, and Garrod and Willis (1997) estimated non-use benefits associated with forest biodiversity.

Mackenzie (1992) used conjoint analysis to estimate the value of six attributes associated with waterfowl hunting trips. The attributes were varied in different combinations, which were described to the respondent as possible hunting trips. The respondent was asked to rate each trip on a scale of 1 to 10. The dependent variable in this experiment was the rating assigned to each trip, and the levels of trip attributes are the explanatory variables in the regression. Using the negative ratio of the coefficient on each non-cost attribute (β_i) divided by the coefficient on the cost (β_{cost}) yields the marginal value or willingness to pay in (9-13):

$$\text{(9-13)}\quad MV_i = -\frac{\beta_i}{\beta_{Cost}}$$

Mackenzie concluded that the conjoint analysis technique can be a useful tool for estimating the value of individual attributes of a public good. This can be important when researchers attempt to value such things as whole ecosystems or forest management options which have multiple attributes. Further, the author concluded that the rating scale approach can account for preference intensities, which would not be accounted for in a discrete choice contingent valuation study. The rating scale model format can also account for respondent indifference or ambivalence. Mackenzie speculated that this ambivalence or indifference manifests itself in the form of non-response to more traditional non-market valuation surveys, leading to bias.

In all the applications above, marginal attribute values were calculated using Equation (9-13). While they are implicit prices, these marginal values may not reflect respondents' maximum willingness to pay for the change in each attribute holding utility constant (compensating variation). Roe et al. (1996) showed that, by including an option that represents the status quo and examining the difference in ratings between the status quo and an alternative, conjoint analysis can yield estimates of compensating variation.

To illustrate conjoint analysis, we report the empirical results of a study that measured marginal passive use values for improving forest health.

2.4.2.3 Application of Conjoint Analysis to the Valuation of Forest Insect Control

Outbreaks of non-native and even native insects and diseases can affect the health and appearance of forests. Pest management actions, such as the decision to spray insecticides will obviously have some desired effect on the target insects, but they may also have adverse consequences on non-target natural resources. Different levels of management will have different levels of impacts and different costs. Even pest management activities themselves involve a wide variety of activities, each with different consequences or effects on attributes. Management to protect one or more attributes of the forest may lead to losses in the levels of other attributes.

Knowing public preferences regarding the trade-offs among various attributes or amenities of healthy forests can be useful to managers who must decide which actions to take on specific forest areas. Obviously a forest near an urban area or in a regional park will merit different considerations than a forest in a National Park. Thus, conjoint analysis may be an appropriate tool to estimate the values of various attributes of management activities to protect or restore forest health. This is because conjoint analysis allows us to examine the disaggregated value of the alternative attributes of different public goods that can be produced from a given forest.

This case study uses conjoint analysis of alternative management programs for three different forest pest situations in the United States. The first, the gypsy moth, is an introduced pest, which has little effect on commercial timber, but does have a high impact on ornamental trees and trees in popular recreation areas. The second scenario involves the western spruce budworm. This insect is native to most fir stands in the western US and has a large impact on commercial timber in the Pacific Northwest. The final scenario is the southern pine beetle, another native insect. This insect has impacts on commercial timber in the Southeast

The three insect infestation scenarios were presented in a questionnaire mailed to 1200 households. The sample was concentrated equally in the three geographic regions most affected by the pests (the Northeastern United States, the Southeastern United States, and the State of Oregon). The same questionnaire (with all three insect scenarios) was sent to each region.

Each forest pest management scenario contained a brief description of the insect, including its area of impact, and the effects of an uncontrolled infestation. Next was a description of three management options, the first of which was "no action." The next option was a moderate level of pest management, and the third was intensive pest management. The three management options were then compared in a table showing the expected effects (e.g., level of attributes for each characteristic) over the next 15 years. These effects are: the area infested, whether the insect was a native pest, changes in commercial timber, and the one-time cost to their household. The attributes used in the survey were derived from Environmental Impact Statements for proposed control programs for each of the scenarios. Levels were based on the expected levels for the alternatives examined in each EIS.

The respondent was asked to rate (on a scale of 1 - 10) the three management options within that scenario. In order to provide sufficient variation in the levels of each attribute to estimate a coefficient on each, 6 versions of the survey were used. These survey versions were identical except for the levels of the attributes in the moderate and intensive management options.

The statistical analysis was conducted using the ordered probit model of the form:

$$(9\text{-}14)\quad RatingsDfference = \beta_1 + \beta_2 Cost + \beta_3 Acres + \beta_4 Timber + \beta_5 Native$$

Table 9-3 contains the coefficients, t-statistics, and log likelihood function for the ordered probit model. The "acres infested" variable and "cost" variable are negative and statistically significant. This indicates that the levels of these variables are inversely related to the individual's rating and utility. The negative sign on the cost variable also indicates internal validity. That is, individuals apparently paid attention to the cost of each option; the higher the cost, the less they preferred it (the lower rating it received). The percent gain in commercial timber is positive and significant, indicating this contributes to the rating.

Table 9-3. Results of Ordered Probit Conjoint Regression

	Coefficient	t-Statistic
Constant	1.781***	13.530
Cost of Program	-0.002481***	-2.782
Acres Protected	-0.001344***	-4.221
Percent Change in Timber Harvest	0.01193***	3.441
Native Insect Dummy	0.20063*	1.680

Note: * and *** indicates significance at the 10% and 1% levels, respectively.

Marginal values for each forest attribute were calculated using the attribute coefficient divided by the cost coefficient. The marginal values can be interpreted as either the loss in benefits from one more unit of a bad, or the willingness to pay to avoid one more unit of a bad. For example, the marginal value on acres infested is $0.54. Thus a household would have a reduction in benefits of 54 cents for each additional acre infested, or it would pay 54 cents an acre to reduce forest infestation by one acre.

Conjoint analysis has the potential to be a useful stated preference valuation tool for environmental valuation of complex public policy programs that involve trade-offs between policy objectives.

3. COMBINING STATED AND REVEALED PREFERENCE METHODS

During the first thirty years of non-market valuation, revealed preference methods such ast TCM and stated preference methods such as CVM, were viewed as two alternative approaches to estimating the value of environmental quality. However, recently several economists have developed hybrid models that combine TCM and CVM data together. There are basically two approaches. The first is sometimes called the trip response model, since it supplements the observed number of trips associated with current quality and price with survey responses regarding "intended" number of trips if quality or price (e.g., travel costs) were changed. These models combine the strengths of both the TCM and CVM. For example, the trip response model is well suited to analyzing the demand for and benefits of management changes that have not occurred at a site. TCM alone would not work, since the management action has not taken place, so there is no way to observe the change in actual trips with the management action. While CVM WTP could be estimated, some economists believe visitors have an easier time stating how they will change their trip-making behavior rather than their value per trip. Studies have shown that responses to intended behavior questions are both reliable and valid (Loomis, 1993).

The second hybrid approach is to combine responses to CVM WTP questions with both actual trip behavior and intended trip behavior. In essence, the individual is asked a minimum of three questions: (a) do you currently visit the site? Those currently visiting the site area asked, (b) if the cost of visiting the site were $Y higher, would you still visit? Finally, (c) would you visit the site if quality were improved by X? (Costs can be left at the current level or raised.) Answers to this series of questions can be analyzed using a variety of panel data estimators available in advanced statistics packages.

By pooling actual and intended behavior, the economist now has the ability to more fully trace out how the demand curve shifts with alternative quality changes (e.g., how visitation is likely to change) as well as how the vertical intercept of the demand curve changes (so that, e.g., WTP changes can be measured) with the same improvements in quality.

4. QUANTIFYING HEALTH BENEFITS OF ENVIRONMENTAL POLICIES

One of the most frequent benefits of policies to reduce air and water pollution is improvements in human health. These health benefits include reductions in illness (i.e., morbidity), injury and, of course, death (i.e., mortality). Determining the health effects from improved environmental policy involves several phases, based on risk assessment, that connect the physical pollution to health benefits. These steps are: (1) reduction in emissions; (2) reduction in concentrations in the environment; (3) reductions in exposure; (4) reduction in effects; (5) valuation of those effects.

Steps 1-4 involve linking pollution to effects on human health. General approaches to linking 1-4 range from high-dose short term studies, often on laboratory animals, to epidemiological studies of human populations. There is of course a great deal of controversy in extrapolating from high does for short periods of time in laboratory animals to low concentrations over lifetimes in humans. The epidemiological approach is typically one of two types: (a) cohort studies that follow a given group of individuals through long periods of time, such as several years; or, (b) daily time series that correlate daily effects on individual health to current and lagged daily exposure levels (Carrothers, et al., 1999). Type (a) studies have the advantage of being long term in nature, so that chronic effects of long term pollution exposure can be measured. The type (b) studies are more likely to capture temporary effects of pollution on human health (e.g., difficulty breathing, sore throat, watery eyes, hospital admissions). The type (b) studies can often take advantage of natural experiments, such as Pope's (1989) study of the effect of a steel mill's emissions on respiratory health of nearby residents. Because the mill opened and closed due to fluctuations in demand, retooling, and a labor strike, there were natural variations in emissions which could be evaluated to determine if there was a corresponding change in illness and hospital admissions. His statistical analysis identified a statistically significant effect.

4.1 Health Effect Metrics

Regardless of the method used, many studies do find a linkage between pollution emissions and human health conditions, including death. There are two different metrics used to quantify the change in risk of death: (a) value of statistical life (VSL); and (b) Quality Adjusted Life Years (QALY) (Carrothers, et al., 1999). The value of statistical life approach is commonly used by government agencies, such as EPA, to calculate the benefits from reducing premature deaths. The VSL measures how much people would pay for small reductions in mortality risk. It is not the value of reducing a known individual's chance of death, but rather reducing the odds that an additional person would die from a given exposed population. For example, a wage hedonic study might show that a worker would have to be paid $500 per year in higher wages to accept an added risk of 1 more death per 10,000 workers. If we divide the $500 annual dollars by the increased risk of death (1/10,000 or 0.0001), we get $5 million as the value of a statistical life. If the individuals being exposed are all about the same age, this approach may be useful. But, some pollutants have a disproportionate effect on the elderly, who have a much shorter remaining life expectancy than the average worker or average person in the population.

To deal with differential life expectancies, policy analysts use the second measure, Quality Adjusted Life Years. This measure is based on remaining life expectancy of a given age cohort of the population. The policy analysis looks at life expectancy with and without the pollution reduction and calculates the additional years gained with the pollution reduction. These additional years or changes in expected survival are then weighted by the health condition of the individual in that age cohort (Carrothers, et al., 1999). For example, extending the life of a 75 year old to 80 years is a gain of five years, but it may provide less utility than extending a 55 year old person's life five years. The ability to enjoy life and the other age-related ailments are argued to put more value on 5 years to the 55-year-old than to the 75-year-old. In the same spirit as the value of a statistical life, these quality adjusted life years can be valued. It is to this issue of how to value improvements in health states or avoiding premature death that we now turn.

4.2 Human Health Valuation Methods

As before, valuation of all benefits in a manner comparable to costs requires that we rely upon a consistent measure, such as willingness to pay or willingness to accept. How much people will pay to reduce their risk of illness or premature death can be ascertained using a variety of methods. Previously we discussed hedonic methods. The hedonic property method estimates how much people will pay to live in safer, cleaner neighborhoods via the higher housing prices necessary. The hedonic wage method estimates how much workers must be compensated to accept higher risk of illness, injury or premature death. The stated preference methods such as contingent valuation and conjoint can also be used to value health related benefits. A CVM example was provided in Section 2.4.1.4 for WTP to reduce exposure to heavy metals.

4.2.1 Cost of Illness Method

Sometimes agencies use cost of illness method to provide a lower bound as to what people will pay to protect their health. As the name implies, the cost of illness method values an injury or illness by the direct medical costs plus the opportunity costs to the individual and society from being incapacitated. These opportunity costs include foregone work hours. Since this method does not account for the disutility of the illness or injury (e.g., the pain and suffering), it understates what individuals would pay to avoid the illness or injury and hence provides a lower bound estimate. Comparison of willingness to pay based methods and cost of illness estimates indicate that the cost of illness approach understates willingness to pay by at least a factor of two, and sometimes a factor of four to six (Dickie, 2000).

While the cost of illness method would seem straightforward, there are many technical difficulties, given the byzantine payment options that are involved with many health insurance plans. It is often difficult to sort out just what the typical individual actually pays, since it depends greatly on what type of health plan coverage that person has. For more details on the cost of illness approach, see Rice, et al. 1985.

4.2.2 Averting Behavior and Defensive Expenditures

Another approach to estimating how much individuals would pay for improved health is to look at the efforts that people take. The first type of behavior is one of avoiding or reducing exposure to the pollutant. This is done by modifying one's behavior, such as remaining indoors during bad pollution days or purchasing air and/or water purifiers (Dickie, 2000). The second type of behavior is to mitigate the effects of the pollution by taking medication or seeking professional medical care (Dickie, 2000). These two measures are closer to the conceptually correct measure of willingness to pay than the cost of illiness approach (Dickie, 2000), but the two averting behavior measures are still frequently lower bounds on an individual's maximum willingness to pay. That is, many times people would pay more than the amount they have to pay for medication, so the price of medication may not reflect their maximum willingness to pay to avoid an illness. However, countering this is the joint production problem in some defensive expenditures. Purchasing bottled water to reduce the risk of pollution may also provide added taste benefits. Purchase of an air conditioner to filter out air pollutants may also cool the room to a more comfortable temperature. Not all of the payment in these two examples is for the pollution reduction, there are other benefits that this method would erroneously attribute to pollution reduction.

This short discussion cannot do justice the conceptual and empirical complexities associated with carrying out a policy analysis involving health benefits. There are, however, several good references to the subject, including Cropper and Freeman (1991), Tolley, et al. (1994) and Viscusi (1993).

5. BENEFIT TRANSFER

For a number of years, economists performing small-scale benefit-cost analyses often were forced by time and budget limitations to use values already reported in the existing literature rather than perform their own studies. (We have previously alluded to this process in Chapter 2.) There are several short-cut valuation methods commonly used by public agencies in such policy analyses. They rely on expert judgment to develop an approximation of the average net willingness to pay. These methods do not require site-specific data collection and statistical analysis other than knowing the number of people affected by a policy or the visitor use levels. The practice of transferring benefit estimates from existing sites to other sites without values has become a subfield called "benefit transfer." Benefit transfer can incorporate unit day value approaches; it can also involve transfer of the entire demand function from a TCM study or a willingness to pay function from a CVM study to a different application. We will now briefly describe these approaches in the remainder of this chapter.

5.1 U.S. Water Resources Council Unit Day Values

The U.S. Water Resources Council's UDVs are the oldest and probably the most well known recreation benefit transfer method. This method relies on expert judgment to develop an approximation of the average net willingness to pay for recreation use. The values selected are considered to be equivalent to consumer surplus, i.e., net of travel cost or price.

The Water Resources Council (1983) guidelines classify recreation into two categories: "general" and "specialized." General recreation includes the majority of activities requiring the development and maintenance of convenient access and developed facilities. Included are most picnicking, tent and trailer camping, flat water boating and warm water fishing, swimming, and small-game hunting. Specialized recreation opportunities are more limited; intensity of use is low; and more skill, knowledge, and appreciation are required. Recreation use of wilderness areas is considered specialized, as are trout fishing, big-game hunting, upland bird and waterfowl hunting, pack trips, whitewater boating, canoeing, and specialized nature photography.

The guidelines recommended a range in value of $9 to $27 per day of specialized recreation in fiscal year 1997. General recreation values are much lower, $2 to $7 per day. Unit day values have been adjusted for changes in the consumer price index (inflation) to the present. For example, the recommended range of specialized recreation values was $2-$6 in 1962, $3-$9 in 1973, $6-$18 in 1982, and increased to $9-27 in 1996. You can update these values to the present by adjusting for changes in the consumer price index.

Unit day values are adjusted to reflect quality considerations that prevail in various regions based on a point score. For example, a reservoir that carries a heavy load of suspended silt or is used beyond capacity would be less desirable and, therefore, of lower point score and hence would be at the lower end of the range of generalized recreation values than one with clear water and fewer users. The availability of alternative opportunities is also considered in assigning values. Higher values are assigned if the population served does not have similar recreation opportunities. On the other hand, if similar recreation opportunities are relatively abundant, lower unit values are assigned, even if a large number of people are expected to use a proposed site. See the U.S. Water Resources Council (1983) or Loomis and Walsh, 1997, for these point tables.

However, there are problems with these guidelines for rating the quality of the recreation experience. The biggest problem is the subjectivity in rating the site. First, a recreation planner rather than the visitor typically does the rating. Depending on how well the preferences or knowledge of the recreation planner matches that of the majority of the visitors, there can be some differences as to the quality rating. Second, the replicability or reliability of the rating is a problem across sites if different people are performing the rating. Also, it may be quite easy to manipulate the rating to value the recreation activity promoted by the agency higher than other types of recreation. In essence, the rating is like the recreation planner trying to guess at the coefficients on quality in a regression.

The USDA Forest Service developed its own unit day values for recreation as part of its implementation of the Resources Planning Act of 1974 (RPA). The values are updated every five years. The first systematic effort was in 1980, with values for forest-based recreation and fish/wildlife based on the existing literature. Walsh et al.'s (1992) comprehensive review of 20 years of TCM and CVM values established a solid empirical foundation for the 1990 RPA values. With minor refinements, the 1990 values were used by the U.S. Forest Service for the 1995 RPA Program

Environmental Impact Statement. Consult the U.S. Forest Service's latest RPA Program Environmental Impact Statement for updates to these values every five years.

5.2 META Analysis Approaches to Benefit Transfer

Two significant advances have improved the practice of transferring values from the existing literature to sites where management or policy decisions must be made, but for which no values nor time to perform a study is available. The first advancement was the application of META analysis to estimate a regression equation relating average consumer surplus per day from past studies to a host of explanatory variables. These variables include: (a) the type of recreation activity; (b) location of the recreation area; and (c) details of the valuation method employed (e.g., TCM, CVM). Pioneered by Walsh et al. (1992) and Smith and Kaoru, these regression equations make possible a more systematic approach to estimating the value per day of a given outdoor recreation activity without performing a site specific study. The analyst simply needs to insert the values of the explanatory variables for the site of interest into the META analysis regression and perform the multiplications to arrive at an estimate of the value per day. Formal evaluation of its accuracy relative to site specific studies has been conducted by Rosenberger and Loomis (2000). Using one META analysis to predict multiple recreation activity values yields estimated benefits per day that are on average 50% to 70% in error compared to the original studies. This error rate might be reduced if recreation activity-specific META analyses are used, instead of one META equation that attempts to predict values for multiple activities. Nonetheless, it is likely that most META analyses would provide only a rough approximation of the value per day compared to a site specific study. However, for some management decisions or policy issues, the approximate nature of the value estimated from a META analysis may be sufficient.

5.3 Benefit Function Transfer

The second major advance in the area of benefit transfer has been the application of benefit-function transfer. Using this technique, the demand function from an existing TCM study or a WTP function from an existing CVM study performed for a similar activity but in a different location is applied to the recreation area under study. Like META analysis, this approach requires that we obtain estimates of the independent variables from sources such as the US Census and distance calculated from maps. These are inserted into the demand or WTP function to estimate value per day. Since the demand or WTP functions are specific to a particular activity or type of recreation resource (e.g., a wilderness area or a reservoir), the approach is likely to be more accurate than META analysis. However, formal comparison of the benefit-function transfer to an original, site-specific study illustrates some degree of error for this improved approach. For transferring TCM demand functions that include site characteristics for a general recreation resource like reservoir recreation (e.g., boating, fishing, swimming) between two similar geographic regions (e.g., within the Midwest), the errors in the consumer surplus per trip average 5-10%, which is likely to be an acceptable range. However, transferring TCM demand functions across the country (e.g., from the Midwest to the West coast), the errors were an unacceptable 250%. See Loomis et al. (1995) for more details.

5.4 Benefit Transfer in Health Valuation

Benefit transfer has also been used in the health benefit evaluation arena as well. A very thorough discussion and example of using benefit transfer methodology to estimate the health benefits from reducing air pollution can be found in Desvousges et al. (1998). In this case, not only are the valuation estimates sometimes transferred, but so is the dose-response function which indicates the effects of pollution on human health. In some cases, the benefit transfer has been from the United States to other countries, including some developing countries. In this case, a simple adjustment is made to the original study benefit estimates by scaling them by income differences between the two countries. See Alberini and Krupnick (1997) for an example of benefit transfer of air pollution reduction from the U.S. to Taiwan.

At present, benefit transfer is an active area of research. The recent assembly of large databases of existing studies should accelerate this. Environment Canada (the Canadian equivalent of EPA), several Federal agencies, and consulting firms have assembled estimates of the value of environmental quality and the associated demand and WTP functions from hundreds of studies. Use of this database is available on the World Wide Web from Environment Canada (for a fee). The database is called EVRI, for Environmental Values Reference Inventory. The growth in number of original studies and the database and benefit transfer protocols may make it easier to estimate recreation benefits when time and budgets do not permit a site-specific study and accuracy is not critical.

6. USE OF VALUATION METHODS BY FEDERAL AND STATE AGENCIES

Revealed and stated preference techniques, such as travel cost and contingent valuation are widely accepted by government agencies for valuing both recreation and other nonmarketed benefits of environmental resources. TCM and CVM have been recommended twice by the federal U.S. Water Resources Council (1979, 1983) under two different Administrations as the two preferred methods for valuing outdoor recreation in Federal benefit cost analyses. Recently, the U.S. Department of Interior (1994) endorsed the environmental valuation techniques discussed in this chapter as acceptable methods for valuation of nonmarketed natural resource damaged by spills, mining waste and other toxic releases.

The U.S. Bureau of Reclamation and National Park Service relied on CVM to value in dollar terms the recreational values and passive use values of alternative hydropower water releases from Glen Canyon Dam into the Grand Canyon. The Montana Department of Fish, Wildlife and Parks relied on a CVM survey of the benefits of viewing and hunting elk when justifying its purchase of additional elk winter range outside of Yellowstone National Park. Several other state fish and game agencies, including Arizona, California, Idaho, Maine, Missouri, Nevada and Oregon, use the Travel Cost Method and Contingent Valuation Methods for valuing fish and wildlife-related recreation.

As can be seen from the discussion of valuing both marketed and nonmarketed natural resources, there is a well developed and conceptually sound basis for derivation of such values. In most cases, these values stem from people's actual behavior in buying products made from natural resources or from visiting recreation sites. The techniques discussed in this chapter simply allow the analyst to measure those values in dollar terms so as to compare benefits to costs of alternative uses of those natural resources. In this way, the public and the manager can determine if resource allocations or management actions are economically efficient or not.

As we have stressed repeatedly, economic efficiency is not the only criterion, but neither should it be ignored.

7. CONCLUSION

Many people associate economic values only with marketed goods; economics, in their minds, refers only to money. Since environmental policy questions almost invariably deal with non-marketed goods, this view would limit the usefulness of economic tools, such as benefit-cost analysis, for analyzing environmental problems. It also means that any economic analysis of these problems would be biased, possibly seriously, by omission of these non-marketed values.

Non-market valuation seeks to make apples-and-oranges comparisons -- the monetary value of hydroelectricity, for instance, versus ecological effects of converting a river into a reservoir-- into comparisons of apples and apples. It is clear, from the activism of those who want to protect environmental resources and from polling information on those not active, that environmental goods are valuable to people. Estimating the dollar value for these goods provides a means of including them in an economic analysis and ensuring that they are not ignored in that analysis.

To some, estimating dollar values to non-marketed goods is unsavory or even offensive. Putting dollar values on these goods somehow cheapens them, by making them finite and substitutable for money. Just as putting a dollar value on emotional values such as friendship seems, well, disgusting; so does saying that protecting an endangered species is worth the same as, for instance, a large commercial aircraft. So, why do we do it?

As we have emphasized in Chapter 3, economic efficiency is unlikely to be (and probably shouldn't be) the only criterion on which a policy decision is made. At the same time, it is useful information to decision-makers and the public. These decisions will be made based on some criteria; would it not be helpful to inform decision-makers of the intensity of people's preferences for the environmental goods (which these methods are intended to measure)? How much are people willing to give up in order to protect the environment? If it can be demonstrated that people will give up a lot for protection, the argument for protection becomes much stronger. Of course, the converse is true too: if people do not put a high value on protection, then the good is less likely to be protected. Politically, that conclusion may be true anyway. Estimating a dollar value to the resource, as a reflection of the intensity of people's preferences, gives helpful information to a decision-maker.

Perhaps because valuation is potentially such a powerful tool, a few caveats should be raised about its use as well. Probably the most important is understanding its limits. Valuation, and welfare analysis in economics in general, is pretty good at measuring marginal changes -- a 20 percent change in water quality, or a 50 percent increase in elk population, for instance. When the decision is non-marginal -- whether a species should continue to exist, for instance -- the valuation exercise is based on extrapolating data beyond any already-observed range. It is often hard to understand the full implications of these non-marginal decisions -- that is, to compare the "with" and "without" scenarios. Might elimination of a particular species lead to the collapse of an ecosystem? What might be the effects of those changes into the future? As we have discussed above, it is necessary to provide information that is as complete as possible in valuation exercises, so that consumers can make truly informed decisions. If we don't fully understand the effects, we are making decisions on shakier footing.

In many cases, though, we have to make decisions, whether or not we fully understand the consequences of our choices. Valuation is a way to provide fuller, even if not complete, information into the decision process. As one coauthor here has frequently quoted her fellow coauthor, valuation is an attempt to take a good whose value, ex ante, is between zero and infinity, and to estimate its value to an order of magnitude. If that approximate estimate is well below the cost of providing or protecting the resource, then serious consideration should go into whether the resource is worth protecting, or whether other resources are of higher priority. On the other hand, if the approximate value is much higher than the cost of protection, then it is highly likely that protection is the efficient thing to do. Doesn't that kind of information sound useful to you?

Chapter 10

Regional Economic Analysis, Input-Output Models and Multipliers

1. NEED FOR REGIONAL ECONOMIC ANALYSIS IN THE POLICY PROCESS

The need to quantify the change in local income and employment related to policy alternatives have been frequently referred to in previous chapters. Addressing the distributional equity criteria often requires some information on changes in local jobs and income (i.e., wages, rents, business income). That is, does a particular policy or alternative affect local economic activity? Is the change in local economic activity relatively small in magnitude compared with the scale of the local economy? (For instance, a reduction of 100 jobs in a county with 10,000 jobs is minor compared to a reduction of 100 jobs in a county with 500 jobs.) More importantly, is much of this change in local employment concentrated in one or two industrial sectors (e.g., mining)? A loss of 100 jobs spread evenly through out the retail, transportation and food service industries may cause fewer problems to a local economy than a loss of 100 jobs concentrated in one particular industry. While the relative weight to give gains and losses in jobs in different industries may often be a value judgment (i.e., does the gain of one health care professional offset the loss of one truck driver?), the decision-makers and the public have a right to know just what is the change in jobs by industry from some public action.

This right to know is exemplified in the National Environmental Policy Act's procedures for preparation of an Environmental Impact Statement (EIS). Specifically, EISs require the analyst to describe the effects of the proposed policy on the human environment, which includes effects on employment opportunities and income. Thus a regional economic analysis is a common feature of most EISs performed by nearly all Federal agencies and many state agencies as well.

It also makes good sense for a decision-maker to pay attention to local income and employmenteffects. While we demonstrated in the benefit-cost chapter that local gains and losses in employment are nearly always transfers of economic activity at the national level (and hence should be omitted from a calculation of national benefits and costs), these changes in local income and employment sometimes take on symbolic proportions in public policy debates. Opponents of an alternative that might restrict large tractor trailer trucks to non-rush hour periods will often raise the specter of massive reductions in local employment. Conversely, proponents of new development will often tout rosy predictions of gains in employment from the proposed action. Local effects will also have disproportionate influence on local decision-makers' attitudes toward a project. There is often much to be gained by an objective regional economic impact analysis of just what the employment effects are likely to be. This helps add an element of realism, and sometimes demonstrates that local employment concerns are a smokescreen for other underlying reasons for local opposition or support.

The objective of this Chapter is to provide the reader with an understanding of the basics of traditional regional economic analysis, the input-output model and multipliers that are derived from it. With this background, the reader will be much better able to judge whether there is even likely to be any sizeable employment effects and, if so, what types of models might be available to calculate the income and employment effects.

2. TRADITIONAL CONCEPTS OF REGIONAL ECONOMIC ANALYSIS

2.1 Economic Linkages and Leakages

The easiest way to grasp the traditional notion of regional economic analysis is to think about how a local economy might react to building a new power plant near this area. By local, we often mean a county or group of counties directly linked to the power plant, due to where the workers would live during construction and operation of the power plant, or where the inputs to run the power plant would come from. The operation of the power plant has both direct effects and indirect effects that we sometimes group under the headings "backward" and "forward linkages" in this local economy.

For example, the operation of the power plant will have direct effects on the local economy as engineers and plant operators are hired (locally and/or from outside the region). These people directly work at the site; hence their wages are the direct effects on the local economy (the importance of the source of these workers is discussed below).

But operating a new power plant requires many additional inputs, such as coal, water for cooling, and heavy equipment. To the extent that the local economy can meet some of this increased demand for inputs, there is an additional stimulation of local industry. This stimulation of local industry to supply inputs is known as the *backward linkages* and gives rise to what are known as *multiplier effects*, to be discussed later in this chapter.

In addition, the availability of additional electrical capacity might induce companies that use large quantities of electricity to move into the region. Hence, additional workers may be employed in these end-using sectors. This stimulation of final demand may create what is called the *forward linkages* to a local economy.

If a local economy is either industrially diversified or has the appropriate land,water, and other factors of production to supply these backward and forward linkages,operation of a power plant may sustain a large amount of local economic activity.

On the other hand, if this area is a rural economy that does not have the diversified economic base to supply most of the inputs or utilize the final outputs, then the spin-off economic effects will be quite small. For example, the coal for the power plant may come from other states, and the electricity generated may go to an urban area far away. Thus, the analyst must be very careful to scrutinize what required inputs are likely to come from local production (and hence represent a backward linkage), and which from outside the local economy as imports, which are referred to as *leakages* from the local economy. While the potential exists for a new local demand for these inputs to make presently nonexistent local production economical, it is important not to let local optimism overwhelm common sense. Often, production of these inputs is subjected to economies of scale so large that a few plants supply the entire national demand for these inputs. One local power plant may not make local production profitable, and much of the power plant's spending will not have much effect on the local economy.

2.2 Indirect Rounds of Related Economic Activity

Once the direct spending and payment of wages occurs in the local economy, there are more ripple-like effects felt in other sectors of the local economy. For example, for every $1000 of electricity produced, the backward linkages suggest that the company requires several hundred dollars' worth of inputs. This stimulates the industrial sectors supplying those inputs. As those support industries are stimulated, they need more inputs as well. That is, if demand for local coal goes up, the local firms that mine coal must hire more labor and buy more raw materials to meet the increased demand. More underground mining requires more mine timbers to support the mineshafts. This stimulates the forest products sector to harvest more trees and lumber mills to process the timbers. Finally, trucks and fuel are needed to transport the mine timbers to the mining firm and the coal to the power plant.

2.3 Role of Imports in Determining the Size of the Total Effect

The amount of stimulation of economic activity depends on how self-sufficient the local economy is in supplying the labor and capital necessary to support the expansion of basic industries. One critical factor worth investigating is the supply of labor inputs to a new or expanding industry.

The demand for more labor arising from a new project is normally met from one of four sources: (a) the hiring of previously unemployed workers in the local economy; (b) hiring of new workers from outside the local area who now relocate their households to this new area; (c) competing away workers from existing jobs in the local economy by offering higher wages in the expanding industry; (d) hiring of temporary workers who maintain their households outside the local economy. Determination of the direct change in local income and employment in each of these cases may be somewhat different. As an approximation in cases a-c, analysts often simply count the number of new jobs created, though this approach ignores the possibility that the three different sources of supply of labor may have different true gains to the local economy. This is often a reasonable simplification to make, unless the circumstances surrounding a particular development project clearly contradict conditions a-c (i.e., it is a short term but large scale construction project, such as a pipeline, that requires specialized labor services not found in the area). When case (d)arises, much of the wage income is lost to the local economy, as leakages are substantial.

2.4 An Example of Stimulation of Economy From Direct Effects and the Resulting Total Effects

To start, we will estimate the annual change in economic activity associated with the opening of a town's first hotel/conference center. Assume that all of the visitors come from outside the local economy (people who live in the town don't need to stay at the hotel). Thus the hotel revenues reflect an injection of new final demand into the local area. To keep this example simple, we will concentrate on just one sector, the retail sector.

Suppose each hotel guest spends $100 a day at the local stores for meals, film, souvenirs, items they forgot to bring (don't you just hate that), and other miscellaneous items. If there are a 1000 guests at this hotel each month, this means $100,000 of new demand on the retail sector of the local economy.

The immediate effect of the first round of spending results in hiring more workers to order more supplies, stock shelves, ring up sales, and wait on customers and staff in the restaurants. In addition, the local owner makes a profit on the additional $100,000 of sales. The local restaurant will buy more bakery products from the local bakery. Of course, the store owners will also buy more inventory from the wholesale distributor (which we will assume is outside the local economy).

The first round effect is illustrated in Table 10-1, where it is assumed that local value added, or retail margin, is 40%. Thus $40,000 of the $100,000 in visitor spending remains in the local economy to pay for local inputs. The composition of the $40,000 in the local economy might include about $6000, or 15%, as a trade service sector payment (e.g., marketing sector), about $2000-$4000 (5-10%) as local transportation charges, with the remaining being direct payments to local producers (e.g., bakeries, laundries, etc.). The other $60,000 is payment to wholesalers and manufactures that are located outside the local economy. Thus the $60,000 is a leakage.

But this $40,000 received as additional income by workers and business owners stimulates another round of spending. That is, part of the $40,000 of additional income will get spent by the recipients of this income. The extent of this second round of spending depends on the propensity to businesses and workers to buy locally (McKean, 1983). If the local workers and business owners buy most of their consumer goods and make most of their investments locally, then a large percentage of this $40,000 will be spent in the local economy. For example, if 70% of the gain in local income is spent or invested locally, the second round change in spending in the local economy is $28,000 ($40,000 * .7).

To keep the example simple, let's assume that the same retail mark-up applies to this spending as well. Therefore, of the $28,000 spent locally, 40% is retained in the local community as local value added. That is, 40% is payment to local factors of production, with the remaining 60% representing imported inputs or purchases from wholesalers outside the local economy. The 40% of $28,000 is $11,200,the gain in local income (wages and business income) realized from this second round of spending.

But of course, this $11,200 gain in local income to workers and business owners in this local economy stimulates further spending by these recipients. If the same pattern of 70% local consumption, with a 40% value added continues, then $7,840 of the $11,200 would be spent locally and $3,136 would become local value added or income. This $3,136 of additional income continues to stimulate spending by the local recipients. If the same pattern continues this $3136 results in approximately $878 of new local income. Following the same pattern as before, the $878 results in about $246 of additional income to workers and business owners. If the same spending patterns continue, this round results in $70 additional income to the local economy. Finally, this $70 results in about another $20 of local income in the final round. Table 10-1 adds all of these rounds up and calculates the total effect of $55,556.

Table 10-1. Illustration of Direct and Indirect Effects and Multiplier

Round	Total Spending	Leakage from Local Economy	Local Value Added (Income)
1	$100,000	$60,000	$40,000
2	$40,000	$28,800	$11,200
3	$11,200	$8,064	$3,136
4	$3,136	$2,258	$878
5	$878	$632	$246
6	$246	$176	$70
7	$70	$50	$20
8	$20	$14	$6
Total			$55,556
Multiplier = Total/First Round			1.3889

The total effect could be calculated from the first round effect by applying the income multiplier derived in Table 10-1. The income multiplier is the ratio of the total income ($55,556) to the direct income or first round effect, $40,000. In this case the income multiplier is 1.3889.This means that each initial $1 of local income generates an additional 39 cents worth of income, for a total effect of $1.39.

This multiplier can be analytically derived by the formula of:

(10-1) Income Multiplier = 1/[1-(%CL*%VAL)]

where

%CL= percentage of income that consumers' spend locally, and

%VAL= local value added by local businesses (i.e., percentage of total business expenditures made in the local economy).

In our example, %CL is 70% and %VAL is 40%. Therefore, the income multiplier is 1/[1-(.7*.4)] = 1.38889, the same number we obtained from adding up the repeated rounds of spending.

These repeated rounds of additional spending generated from the initial increase in final demand is another way to define the multiplier. If we define the initial increase in final demand as the direct effect and the subsequent rounds of spending as the indirect effects, then the multiplier can also be defined as:

(10-2) Multiplier = (Direct + Indirect Effects)/Direct Effect

The analytic formula for the income multiplier highlights the importance of local consumption and leakages in the multiplier. The greater the local percentage consumption or the more self-sufficient local industry is in meeting local demands, the greater the income multiplier. In other words, a self-contained community will receive the full benefits of local changes in income and suffer the full effects of local losses.

Generally, businesses in remote rural towns import a high percentage of their inputs, and many consumers make large purchases out of town. Thus value added or income multipliers of 1.25 to 1.75 are common in these rural areas. As Coppedge (1977:4) states with regards to income multipliers, "Most estimates would fall between 1 and 2. ... A multiplier that exceeds 2 should be subjected to critical review before acceptance or use in further analysis". Of course a larger geographic area and size of the economy covered increases the multiplier. Thus a multiplier for "Tiny Town", Iowa might be 1.3, but for Los Angeles, California, it might be 1.75, and for California as a whole it may be 2.5.

3. INPUT-OUTPUT MODELS: A SIMPLE MODEL OF A LOCAL ECONOMY

Now that you have an intuitive understanding of the income multiplier and regional economic linkages, it is time to more formally develop the analytical methods used to calculate the multipliers for all sectors of the economy. To start with, we will consider two views of a given local industry. The first view is based on their output, the second view based on their use of inputs. These are the foundation of input-output models used to derive multiplers in complex economies. Input-output models are the heart of more sophisticated regional economic models such as Social Accounting Matrix (SAMs) and computable general equilbrium (CGE) models.

The output of a given industry can be used in one of five general ways. A given industry's output can be directly consumed by households (C), it can be used for an investment in capital goods (I), it can be consumed by government agencies (G), it can be used as an input by other local industries (x_j), or it can be exported to anyone of these uses outside the local economy (E). Equation (10-3) presents this relationship for the output (X) of industry i:

$$(10\text{-}3)\ \sum_{i=1}^{N} X_i = \sum_{j=1}^{J} x_{ij} + (C_i + I_i + G_i + E_i)$$

where x_{ij} is the amount of industry i's output used as an input (or intermediate good) by industry j.

The amount $C_i + I_i + G_i + E_i$ is often referred to as the *final demands*, to distinguish it from x_{ij}, which are *intermediate demands* of the processing or producing sectors. Of course the other way to view industry i is from its use of inputs. For example, industry i employs labor (and pays wages, W), uses land (L), capital (K), raw and semi-finished materials produced by other industries (x_{ij}), and management inputs (which are paid a profit, P). Additionally, some of the inputs might be imports from other industries outside the region (M). What many regional economic models do is to trace the factor payments to both local owners of inputs and importers of inputs. The payment view of equation (10-3) is given in equation (10-4): the total outlay (X) of industry i equals:

$$(10\text{-}4)\ \sum_{i=1}^{N} X_i = \sum_{j=1}^{J} x_{ij} + (W_i + L_i + K_i + P_i) + M_i$$

where

W_i is wages paid to workers in industry i

L_i is rental payment on land used in industry i

K_i is rental or interest paid for use of capital in industry i

P_i is the profit or return to management in industry i

M_i is the payments to imported inputs (labor, materials, capital, management) from outside the region.

As you can see, the left hand side variable in equations (10-3) and (10-4) are the same. This is because the value of the output produced by an industry (10-3) must equal the payments to all of the factors of production used, including profits (10-4). That is, the value of the output must equal the total cost of production plus profit.

Given the equality of factor payments and value of output, Gross Regional Income (what the local economy receives) equals Gross Regional Product (the value the local economy produces; Richardson, 1972: 20)). Specifically, since the right hand side of equations (10-3) and (10-4) both equal X_i, then:

(10-5)

$$\sum_{i=1}^{N} x_{ij} + (W_i + L_i + K_i + P_i) = \sum_{i=1}^{N} x_{ij} + (C_i + I_i + G_i) + (E_i - M_i)$$

Gross Regional Income = Gross Regional Output

The relationship shown in equation (10-5) is the building block of the transactions table that underlies many Input-Output models used to describe entire economies and calculate multipliers for all of their sectors. The Transactions Table is made up of three components: a processing or producing sector (the interindustry part, or x_{ij}s); the payments sector, and the final demand sector (C_i+I_i+G_i+E_i-M_i). Thus any economy can be viewed as having a core producing sector that takes inputs and satisfies final demands.

Payments to the factors of production (including taxes) used by the processing sector allow the households and the government to finance their final demands for goods and services. In this sense the elementary circular flow in any economic system is represented in the Transactions Table. Table 10-2 provides a simplified Transaction Table that will serve as the basis for our discussion.

Table 10-2. A three sector transactions table

	Sales to->	Processing Sector			Final Demand	Total
		X1	X2	X3		
Purchases From:						
	X1	2.00	4.50	0.40	3.10	10.00
	X2	4.00	12.00	2.00	32.00	50.00
	X3	0.40	6.00	3.60	30.00	40.00
Payments To:						
	Labor	1.60	20.00	15.00	4.00	40.60
	Other local inputs	0.80	3.00	10.00	1.00	14.80
Imports		1.20	4.00	9.00	1.00	15.20
Total Outlay		10.00	50.00	40.00	70.10	170.00

This Transaction Table provides a "double entry" bookkeeping view of this simplified local economy. Reading across the first row, industry #1 (X1) sells itself $2 worth of its own output (that is, uses some of its own output as an input to produce more output). Industry #1 also sells $4.50 worth of its output to industry #2 and just 40 cents worth of output to industry #3. Industry #1 also sells $3.10 worth of its output directly to final consumers (e.g., consuming households, government, or export out of the region). Thus industry #1's sales of output is $10.00.

Now, reading down the first column shows a factor payments view of industry #1. Specifically, it shows what industry #1 must buy or pay out to get the inputs it needs for its own production. Here, industry #1 buys $2 worth of its own output, $4 worth of industry #2's output, and just 40 cents from industry #3. In addition to these materials, industry #1 buys $1.60 worth of labor (i.e., pays wages and salaries to workers of $1.60), uses 80 cents worth of other local inputs (land and building rent, local taxes, profits to local owners, etc.) and imports $1.20 worth of other inputs not available from the local economy. Thus, to produce $10 in sales of total output (reading across the first row), it requires $10 in total outlays (the first column). Thus there is a balance for X1, and there is balance for every other processing sector as well.

The transactions table illustrates the economic interdependencies among processing sectors themselves and the processing sectors and the rest of the economy. The magnitude of these economic interdependencies determines the size of the income multiplier in each sector and, hence, how much spin-off economic effect there will be if a new factory or plant is allowed to enter the town.

To estimate the multipliers from the transactions table, two additional steps are necessary. First, the policy analyst must construct the direct coefficients table using the information in the transactions table. The number in this direct coefficients table represent the dollar amount of other inputs required from a given processing sector to produce one dollar's worth of output. This is the first round effect discussed earlier. Given this definition of direct coefficients, it is easy to see how they are calculated from the transactions table: the direct coefficients are the individual matrix elements from the transactions table divided by the column total (i.e. total outlay for that industry).

For example, the technical coefficient for industry #1 buying from itself (denoted X_{11}) is $2/$10 or 0.20. For industry #1 buying from industry #2, the coefficient X_{21} is $4/$10 or 0.40, and so forth. These technical coefficients show the first round economic interdependence between sectors on a per dollar basis. That is, for industry #1 to produce an additional $1 of output, it must purchase 40 cents' worth of industry #2's output. This technical coefficient relating what industry #1 needs from industry #2 is usually labeled a_{21}.

But of course, for industry #2 to produce more output, it must purchase inputs from other industries as well. Using the technical coefficient matrix shown below in Table 10-3, industry #2 must buy 5 cents of output from industry #3. This is additional stimulation to industry #3 over and above the direct stimulation received by purchases from industry #1 itself (4 cents) is the second round effect.

Table 10-3. Technical coefficient matrix

		Processing Sectors		
Sales To ->		X1	X2	X3
Purchases From	X1	0.20	0.09	0.01
	X2	0.40	0.24	0.05
	X3	0.04	0.12	0.09

Similar effects occur in the third round as input suppliers to industry #3 are stimulated in response to industry #3's attempt to supply more input to industry #2. To calculate the multipliers, the information in the technical coefficient matrix is used along with the relationship in the transactions table. Specifically, that output of X_1 is equal to:

$$(10\text{-}6)\ X_1 = a_{11}{*}X_{11} + a_{12}{*}X_{12} + a_{13}{*}X_{13} + Y_1$$

where Y equals the final demands of consumers, government, investment, and exports. Thus equation 10-6 is similar to equation 10-3 earlier. The key addition is that, with the technical coefficients (a_{11},a_{12}, and a_{13}), one can calculate the multipliers from a series of equations like 10-6.

We said series of equations, because, of course, there is one equation like 10-6 for each processing sector. In our example there are three such equations. In real I-O models there can be more than a hundred processing or manufacturing sectors, hence more than a hundred equations like (10-6). To make calculation of the multipliers more direct, the series of equations are algebraically rearranged and put in matrix form as:

$$(10\text{-}7)\ \mathbf{Y} = \mathbf{X} - \mathrm{A}\mathbf{X}$$

where **Y** is a matrix of final demands and **X** is a matrix of total output of each processing sector. A is the technical coefficient matrix (i.e., Table 10-3). To begin to solve equation 10-7 it is rearranged as:

$$(10\text{-}8)\ \mathbf{Y} = (\mathrm{I} - \mathrm{A})\ \mathbf{X}$$

where I is an identity matrix (i.e., a square matrix with the main diagonal elements equaling 1 and the off-diagonal elements equaling 0). To solve for the multiplier that measures the direct plus indirect output associated with an expansion of final demand for X, (10-8) is rearranged one last time:

$$(10\text{-}9)\ \mathbf{X} = (\mathrm{I}\text{-}\mathrm{A})^{-1} * \mathbf{Y},$$

where -1 as a superscript denotes a complex matrix mathematical operation called the inverse. In this case, the $(I\text{-}A)^{-1}$ is called the Leontief inverse, for the developer of the input-output model. Solving this inverse is equivalent to computing all of the iterations in rounds of spending that would occur (as we did in Table 10-1) in response to a one dollar increase in final demand. Table 10-4 illustrates these coefficients for our example.

Table 10-4. Direct and indirect coefficients per dollar of final demand

		Processing Sector	
Purchases From	X1	X2	X3
X1	1.33	0.16	0.02
X2	0.77	1.41	0.08
X3	0.15	0.19	1.11
Output Multipliers	2.25	1.76	1.21

The interpretation of these multipliers is that, for every one dollar increase in direct final demand for sector X_1's output, there will be a \$2.25 total increase in X_1's output. This occurs because of the linkages between X_1 and the other sectors of the economy (here, X_2 and X_3). Due to the smaller linkages of X_3 to the other sectors of the economy, the multiplier for X_3 is smaller.

3.1 Distinguishing between Type I and Type II Multipliers

The multipliers in Table 10-4 meet the definition of a type I multiplier. A type I multiplier is defined, as in equation (10-2) above, as the ratio of direct plus indirect effects to the direct. These indirect effects in Tables 10-1 and 10-4 only reflect the indirect effects on inter-related processing or manufacturing sectors of the economy.

However, as discussed earlier, the additional factor payments (wages, profits, etc.) become income to households. Naturally, households will spend a large portion of this additional income. As was seen earlier, this sets off additional rounds of spending in the economy. This additional effect of additional consumer demand by households associated with the direct and indirect effects is called an induced effect. If we include households directly in our transactions table, we will pick-up this induced consumption effect in our estimate of the multiplier, and, hence, we will obtain the type II multiplier. The type II multiplier is defined as:

(10-10) Type II Multiplier = (Direct + Indirect + Induced Effects)/Direct Effect

Not surprisingly, the type II multiplier is larger than the type I. Often the type II multiplier is much (25 to 50%) larger than the type I multiplier. The type II multiplier is often the appropriate one to use to capture the full long run effect of a permanent change in final demand in a local economy in which there will be household response.

Calculation of the type II multiplier can be done within the Input-Output model framework. The key change is to expand the view of the processing sector portion of the transactions table to include households as a "processing sector." This is sometime referred to by analysts as making households "endogenous" in the model. That is, household consumption demand is partly determined by the interactions among the sectors rather than being solely treated as a predetermined constant outside of (i.e., exogenous to) the processing sectors.

3.2 Social Accounting Matrix (SAM)

The SAM is an expansion of the basic I-O transactions table to include non-industrial financial flows, such as factor and institution exports and inter-institutional transfers (MIG, 1997:136). Factor and institution exports are payments made by firms outside the study area to workers or individuals living inside the study area. A factor export would be stock dividends or interest income received from firms or banks located outside the study area. Another non-industrial financial flow is retirement income. These have been an increasingly important source of income in many areas of the U.S. economy. In some cases it is the environmental attributes of the area that draw these owners of capital and retirees there. Sometimes stimulating the industrial sector of an economy to the detriment of the natural environment may result in departures of owners of capital and retirees, resulting in a net reduction to the economy (Power, 1996). An institutional export would be wages received by a worker that commutes to a job outside the study area, but lives in the study area. Inter-institutional transfers involve payments from one level of government (e.g., federal or state) to another level of government (e.g., county). For example, Federal Payments in Lieu of Taxes by the Federal government to compensate local governments for inability to collect property taxes on federal lands are important inter-institutional transfers in many public land counties in the western U.S.

While these distinctions may seem rather esoteric, one of the defining features of what has been called the new economy is the importance of these three sources of "extra-regional" income. A study by Waters, et al. (1999:275), of the Oregon economy found, using a SAM rather than I-O model, that this outside regional income "..emerges as the most important generator of jobs in the Oregon economy, providing the source of almost 20% of Oregon jobs in 1993." This figure is several times larger than logging's contribution in Oregon.

Another advantage of a SAM is the ability to disaggregate households by income level so as to better evaluate distributional effects of policies. For a discussion of SAM, see Pyatt and Round (1985).

3.3 Income Multipliers

Much as the business or output multipliers were determined above, the income multiplier can be analogously determined in the Input-Output model. The type I and type II income multipliers can be defined similarly to equation (10-2) and (10-10) i.e., the type I income multiplier is the ratio of the direct plus indirect change in income to the direct change in income.

The key difference between output multipliers and income multipliers is that the income multiplier applies just to the direct change in income, not to the direct change in dollar output or sales. A frequent mistake is to multiply an increase in $10 million in final demand by an income multiplier (of say, 1.5) and conclude that $15 million in local income will be derived. The fallacy here is that much of the $10 million increase in final demand will be met by importing more inputs into the local economy. The direct gain to the local economy is only the local value added. It is this local value added which represents the direct or first round income effects. Therefore, it is these first round income effects that should be multiplied by the income multiplier.

3.4 Employment Multipliers

Besides measuring the change in local income, often the policy analyst and decision maker are interested in the change in local employment associated with the direct change in final demand in the local economy. This requires that one additional piece of information be developed to be used along with the Input-Output model. That piece of information is the number of workers per dollar of output delivered to final demand. For example, each $10,000 of microchip production might require 0.035 workers.

This is the direct employment coefficient that is in some respect analogous to the direct output or technical coefficients presented above in Table 10-3. Of course the policy analyst and decision maker are also interested in the indirect and total employment effects. To calculate this requires multiplying the matrix of direct employment coefficients by the direct plus indirect coefficient matrix (e.g., Table 10-4). If the form of Table 10-4 is used with households exogenous, the policy analyst obtains type I employment multipliers. If households are endogenous, then type II employment multipliers result. The type II employment multipliers can be about 1.5 to 3 times larger than the direct employment effects alone (McKean, 1981: 39-40).

4. ASSUMPTIONS OF INPUT-OUTPUT MODELS

The ability of Input-Output models to concisely summarize the interrelationships between industries themselves and other segments of an economy (households, government, etc.) requires several simplifying assumptions. One way to summarize these assumptions is in three words: *linear additive production*. These three terms refer to the first assumption of Input-Output models:

4.1 Constant Returns to Scale

Each industry's and firm's production function is a fixed and constant relationship between inputs used and outputs produced. For example, using 10% more inputs yields exactly 10% more output. To obtain 10% more output requires that all the inputs be increased by 10% because the production recipe or mix that translates inputs to outputs is exactly the same at 10% more output and 100% more output. In the short run time period of a few years, this assumption may be plausible. The firms and industry have a given set of machines and other capital that limits the degree of deviation from the existing production mix.

However, in the long run, this assumption rules out several common types of long run economic behavior of firms. The fixed relationship between inputs and outputs rules out any economies (or diseconomies) of scale when producing large increases in output to meet major increases in final demand. Again, one often observes, as firms expand in the long run, that they can take advantage of economies of scale. Related to this assumption is the lack of substitution possibilities even as the level of production is changed.

4.2 Single Industry, One Output

Traditional industry I-O models assume that each commodity or group of commodities is produced by a single industry in which each firm in the industry produces only that output. More recent I-O model structures no longer requires such an assumption and allows for multiproduct industries (Alward, et al, 1985). However, all models still require that each firm uses the same mix of inputs to produce that output. When product groups or industrial sectors are broadly defined, this assumption is at best an approximation.

The more narrowly defined the industry, the more likely some firms will: (a) also produce other related goods; (b) use production technologies that are somewhat different from one another.

4.3 No Supply Constraints

I-O models assume that there are no constraints on a firm or industry in expanding output to meet increased demand. That is, there are no bottlenecks in terms of land, labor, or capital that would interfere with industry expansion (Richardson, 1972: 27). This assumption may sometimes be violated in the short run, when firms are unable to quickly locate the additional specialized labor or capital goods necessary to expand production. In some urban areas, the availability of land or water may a binding constraint. That is, while a firm may have identified the land needed to accommodate the proposed manufacturing plant, the indirect and induced effects predicted by the IO model may require additional land or water that is not available locally.

When these assumptions are not be fully met, newly developed computable general equilibrium (CGE) models may more realistically predict the regional economic impacts to a local economy. As usual, these more flexible and realistic models are more complex to set up, as they involve estimating demand and supply relationships for major sectors and outputs in the economy. Many policy analyses continue to rely on simple Input-Output models, since they often believe that the cost of additional complexity outweighs the benefits of greater precision. When possible, the policy analyst can qualify the results of Input-Output computations when communicating these results to decision makers. For example, a policy analyst familiar with a local economy can indicate whether bottlenecks hindering the ability of local firms to expand might result in only part of the increase in final demands being satisfied locally and the rest being imported (even though the simple model would indicate full local expansion of production).

5. DEVELOPMENT OF INPUT-OUTPUT MODELS: THE SURVEY BASED APPROACH

The data necessary to construct a Transactions Table for sub-state areas often involve a combination of existing secondary data from government agencies and primary data collected from the industries themselves. Often both mail questionnaires and in-person surveys are used to obtain the detailed information on what inputs an industry uses, where those inputs are obtained, and the rate of output associated with those inputs. Since total outlay must equal total output, the various sources of information must be reconciled so that the overall transactions table is balanced.

Development of an Input-Output model from primary data is a substantial undertaking and requires extremely detailed knowledge of regional economics. Typically most policy analysts, consultants, or government agencies do not build primary data Input-Output models themselves. Rather, they use existing Input-Output models or contract out to regional economic specialists to develop such models.

6. USE OF EXISTING INPUT-OUTPUT MODELS

Many states have developed primary data Input-Output models for their states as a whole. For example, Utah has developed its own Input-Output model for the entire state (Bradley and Gander, 1968). Other states like Colorado, Oregon, Washington, and Wyoming have developed Input-Output models for various sub-state regions of their economies. Whenever these models match the geographic scope of interest in a policy analysis, it is wise to use these existing input-output models.

6.1 RIMS Multipliers

A consistent set of state-level output, income, and employment multipliers are available for each state in the U.S. from the U.S. Department of Commerce (1986). These multipliers are known as RIMS, for Regional Input-Output Modelling System. These multipliers are based on the United States national averages production technology. However, the multipliers do reflect the presence or absence of specific industrial sectors in a state's economy.

The RIMS income multipliers are slightly different than discussed above. These income multipliers reflect the combined effect of state value added and a type I multiplier. Palmer (1976:19) calls such a number a "response coefficient" rather than a multiplier, although RIMS calls them earnings multipliers (U.S. Department of Commerce, 1997). An *income* multiplier converts the direct change in *income* to the total change in income. The response coefficient converts the changes in *sales* volume or total gross output to income. Therefore, the RIMS income coefficient/multiplier can be used directly against gross output or sales volume, since the state value added has already been accounted for in determining the size of the multiplier. Of course this makes the resulting RIMS multiplier smaller. To illustrate the differences, recall that the income multiplier in Table 10-1 was 1.3889 and was multiplied not by the initial $100,000 spending, but rather it was multiplied only by the initial or direct income of $40,000. Total income was therefore $55,556. The equivalent calculation using a RIMS multiplier would be .55556 time the $100,000 spending for the same total income of $55,556. Essentially the RIMS multiplier of .55556 reflects the fact that only $40,000 of the initial $100,000 spending would be direct income, with the other $60,000 being leakage outside the area. So if you take $40,000/100,000 time 1.3889, you will end up with the RIMS multiplier of .55556. Thus, most of the RIMS income multipliers are between are decimals, usually ranging from .45 to .78. However, as illustrated above you should not be fooled by this smaller number. It contains the same full indirect effects of a more standard income multiplier (Palmer, 1983).

Often there will not be a perfect match between the geographic region of interest and the available state level Input-Output models. For example, if one were performing an analysis of building a power plant in northern Michigan, the State of Michigan Input-Output model would not be very appropriate. The reason is that the state of Michigan as a whole is much more self-sufficient than would be the remote upper peninsula economy of northern Michigan. Southern Michigan is a major manufacturing, wholesaling, and distribution center for the north central United States and hence has a relatively high income and employment multipliers. This is contrasted with the small towns of in northern Michigan that would be the main support centers for construction and operation of the power plant. These small towns import much of their retail goods, food, etc.. Thus, the leakages from these economies are quite high. As a result, the income and employment multiplier would be much smaller for these areas than for the State of Michigan as a whole, and the use of the State of Michigan Input-Output multipliers would greatly overstate the economic effects of power plant construction on the northern Michigan economy. If, however, the geographic scope of interest was the entire state of Michigan, then the state level Input-Output model would be appropriate.

To address this problem and provide updated estimates, recent versions of the RIMS multipliers, now called RIMS II (U.S. Department of Commerce, 1997), provide the capability to generate multipliers for any county or group of counties in a region. Industrial detail can be provided for 38 industry aggregations, or up to 490 individual industries (if they are present in that region). The U.S. Department of Commerce will provide custom analysis to generate the output, earnings, and employment multipliers for any user-specified region. The multipliers are based on the Bureau of Economic Analysis' 1992 I-O table. The analysis is performed on a cost reimbursable basis, but at the time of writing (2001) it was less than $300 per region. More information about the RIMS II multipliers can be obtained at U.S. Department of Commerce's web site, under Bureau of Economic Analysis (http://www.bea.doc.gov/bea/uguide.htm#_1_32). In addition, the manual is available at:
http://www.bea.doc.gov/bea/ARTICLES/REGIONAL/PERSINC/Meth/rims2.pdf

6.2 Tailoring an IO Model to a Sub-State Local Economy: The IMPLAN Model

The need for Input-Output models that reflected the economic interdependence of rural counties that surround most National Forests has led the U.S. Forest Service (Alward and Palmer (1983)) to develop a modular Input-Output model that could be adapted down to the individual county level, for any county in the U.S. The resulting IMPLAN model consists of two major parts: (1) a national-level technology matrix and (2) estimates of sectoral activity for final demand, final payments, gross output, and employment for each county. The database represents 1977 through 1990 county-level economic activity for 528 sectors or industries. The national technology matrix denotes sectoral (industry) production functions and is utilized to estimate local purchases and sales. This 528-sector, gross domestic product-based model was derived from the Commerce Department's national input-output studies.

In essence, the information on what sectors are present in what counties is used to construct a transactions table with only the processing sectors that are present in the local economy. Everything else is assumed to be imported. The national average technology is used to develop the direct coefficients for the sectors that are present locally. This is an added assumption of IMPLAN: that local firms and industries utilize the national average production technology and are on average as efficient at turning inputs into outputs as the typical U.S. firm. Comparisons with locally derived survey-based Input-Output models have shown that IMPLAN may overstate the local multipliers slightly (Radtke, et al., 1985).

The strong feature of IMPLAN is that the policy analyst can tailor or customize an Input-Output model for any area (zip code, county or group of counties, state or groups of states) affected by a proposed policy alternative. In the example below, a impact area was defined as two zip codes. The IMPLAN database is updated by the Minnesota IMPLAN Group as the Bureau of Economic Analysis collects new data and re-estimates the National Input-Output model. This is the same database used by RIMS II multipliers.

The structure of IMPLAN also reflects recent advances in design of I-O models. To avoid the homogeneous, single-product industry assumption of traditional square inter-industry I-O models, an industry-by-commodity format is being adopted in IMPLAN. As described by Alward et al. (1985), "In this format, industry demand for commodities is given by the 'use' matrix and industry production of commodities is provided by the 'make' matrix." The make matrix viewpoint allows for a distinction between local commodity production that is for domestic consumption and local production for export. Another extension is to a Social Accounting Matrix (SAM) approach that allows for tracing of incomes received not just by location of employment but also location of residence and by use of that income in the consumption process. More importantly, the SAM approach allows for explicit incorporation of non-production flows of income, such as governmental grants-in-aid from Federal to local governments. See Alward et al. (Appendix A), 1992, for a detailed discussion of the SAM approach.

While IMPLAN was originally developed for mainframe computers, micro-computer versions are now available from the Minnesota IMPLAN Group in Stillwater, Minnesota. The current version of IMPLAN runs in a Windows environment. The Minnesota IMPLAN Group not only sells the software but provides training. While IMPLAN was developed by the U.S. Forest Service, it is now widely used by many state and federal agencies, universities, and private consulting firms.

7. EXAMPLE APPLICATION OF IO MODELS AND THE IMPORTANCE OF STATISTICALLY ESTIMATING THE INITIAL DIRECT EFFECTS

The purpose of this case study is to illustrate calculation of the initial or direct change in final demand and the corresponding IO estimates of jobs for a small local economy. The example evaluates the relative effects that population growth and possible global warming would have on visitation to a national park and hence employment in its gateway community. While climate change is potentially important, population growth is also likely to be a major factor in determining economic impacts. Both climate and population will influence visitation to the national park. The change in visitor expenditures is the initial change in direct final demand that will be used by the IO model to predict the total effect on employment in the gateway community's economy.

This example also illustrates the role of policy analysis at the local level: while a small local area will have little effect on state population growth or on global climate change, it can plan to partially mitigate the effects via appropriate adaptations. For example, if population increases and global warming would result in large increases in visitation, and hence demand for visitor services, this demand will stimulate the local economy. This stimulation will create demands for additional employee housing and the need for improved transportation infrastructure to allow for movement of the additional employees and visitors around town. In order to simulate a useful planning horizon, this example focuses on determining the 2020 climatic and population effects on visitation. In our example, this would give Rocky Mountain National Park in Colorado, and its primary gateway community, the Estes Park area, adequate time to prepare for the predicted level of growth.

The analysis of the climate change driver includes the use of monthly changes in precipitation and temperature under higher levels of greenhouse gas (GHG) concentrations. A global circulation model (GCM) predicts that these higher levels of GHG will result in global climate more than 4° F warmer than the historical baseline period and a drier overall climate. The key economic impact question is how this change in climate might affect visitation to Rocky Mountain National Park and the local economy of Estes Park.

Calculation of the initial or direct effects of some policy is often the most challenging and least mechanical part of a regional economic analysis. This may be where advocates and opponents of some proposed policy may markedly differ, with one group suggesting large initial job gains and another very small job gains. The accuracy of many regional economic analyses could be substantially improved by putting more effort into modeling the direct change in final demand that will be used as the initial driver in the IO model. This example will illustrate using multiple regression model estimates (see Chapter 6) to drive the IO analysis. The regression model will be used to test the hypotheses that the changes in Colorado population create a statistically significant effect on employment in the Estes Park area, and that there is no statistically significant increase related to climate change. To reject this hypothesis implies that climatic changes, as predicted by the GCM model, also have a statistically significant impact on employment in Estes Park, separate from the effect of population increase.

7.1 Tailoring IMPLAN Model to Zip Code Level

Estes Park is a small economy that is characterized by its dependence on tourist visitation. This type of economy typically has most sectors producing goods and services that are consumed mainly by visitors to the economy rather than by other industries in the economy. Visitation effects on the final demand sectors are thus felt primarily through increases in demand for particular industries' products, such as lodging, recreation, and food.

The IMPLAN IO model was used as the primary basis for this analysis. Given the small economy of Estes Park, zip code level data is best suited for this analysis. As it is often important to do, the original IMPLAN zip code data were updated with several local, state, and national sources. The model was then calibrated to include households and local government as endogenous sectors, which should accurately reflect spending cycles in such an economy and yield type II multiplier effects.

Because visitor expenditure profiles did not exist specifically for Rocky Mountain National Park (RMNP), visitor expenditure data were gathered from two other national parks chosen for their similarity to Estes Park. Yosemite National Park was chosen because, like Estes Park, it is situated near a large metro area; Yosemite's total visitors' per capita spending was thus used as a benchmark. Grand Teton National Park (GTNP) has a single major gateway playing a similar role to the role that Estes Park provides for RMNP. This symmetry suggests that spending proportions for GTNP were likely to be broadly representative of expenditure shares of RMNP visitors in Estes Park. This approximation of changes in spending per unit (per visitor in this case, but per employee, per square foot of retail space, etc.), is another common hurdle in accurate use of IO models in applied policy analyses.

From the combined data gathered from these two parks, an estimate of $68 per visitor of local expenditures is divided among six sectors. This total spending must be disaggregated out to match IMPLAN economic sectors. These sectors are lodging ($27), eating/drinking establishments ($12), travel ($11), equipment/miscellaneous retail ($11), recreation services ($4), and food stores ($3). Since visitation drives the final demand in these sectors, the spending by typical visitors will produce impact estimates on affected sectors in the areas of output, employment and income.

To calculate the change in final demand, we not only need expenditures per visit, but also how total visitation changes with climate and population. Thus, we model total visitation (V_i) to Rocky Mountain National Park for month *i* as a function of population and several climate and variables:

(10-11) $$V_i = \beta_0 + \beta_1 S_i + \beta_2 T_i + \beta_3 P_i + \beta_4 Pop_i + \beta_5 SVDV_i + e_i$$

where S_i represents average snow depth for month *i*, T_i represents average maximum temperature for month *i*, P_i represents total precipitation for month *i*, Pop_i represents average monthly population for 12 counties along Colorado's Front Range, $SVDV_i$ represents a dummy variable for school vacation months ($SVDV = 1$ for July and August), and e_i represents the normally-distributed disturbance term.

Visitation data were provided by Rocky Mountain National Park administration for the years 1987-99. Population data and projections for the Front Range counties of Colorado were gathered from the state demographer's office. The regression coefficients on the independent variables were generally statistically significant, and the explanatory power of the peak season model (the R^2) is quite good at 87%.

The regression model was first applied allowing only for population change. Then it was used with the predicted GCM temperature and precipitation along with the population change; the results provide estimates of monthly visitation with and without climate change. The results of this estimation are provided below in Table ?-5. The standard error of the regression estimate is used to compute confidence intervals around the forecast that will allow calculation of confidence intervals around the Estes Park job estimates from the IMPLAN IO model.

Table10- 5. Visitation and total jobs in Estes Park with and without climate change

Scenarios			Current Baseline Regression	Forecast/ No Climate Change	Forecast/ GCM
Total Season Visitation			2,536,303	4,885,404	5,221,941
Total Season Jobs			4,844	9,330	9,973
Monthly Standard Error			71,808	116,155	117,683
Confidence Intervals	Minimum	Visits	1,674,612	3,491,545	3,809,750
	Maximum	Visits	3,397,993	6,279,263	6,634,132
	Minimum	Jobs	3,198	7,068	12,670
	Maximum	Jobs	6,490	11,592	7,276

As can be seen from Table 10-5, the confidence intervals for current baseline jobs and 2020 population jobs do not overlap, signifying that they are statistically different from one another. Therefore, Colorado population growth between the two periods is expected to have a significant effect on visitor expenditures, local economic activity, and total jobs in Estes Park. The confidence intervals for the combined 2020 population and the climate change scenario jobs do overlap the population-only scenario, implying that the changes in jobs due to the combined effect of population and climate change were statistically indistinguishable from those due to population alone. This example illustrates the importance of the confidence intervals in assessing significance of different changes in initial final demand in an input-output model. In this case, the more certain population growth effects on visitation suggest planning for increases in employee housing and improved transportation infrastructure even if global warming fails to materialize.

8. PUTTING LOCAL CHANGES IN INCOME AND EMPLOYMENT IN PERSPECTIVE

While, from a regional economic impact standpoint, it is useful to understand how a policy action affects local income and employment, it is important for the policy analyst to keep in mind these are just local effects. Gains of employment in one county being analyzed are usually offset by reductions of employment elsewhere in other counties outside the impact area, if there is full employment. (Full employment involves an unemployment rate around 3-5 percent. Most of this unemployment is frictional, i.e., people temporarily unemployed while they look for better jobs or move to a new job.) Thus we caution you from generalizing that some policy action will "create" new jobs or result in job losses. A policy may have this result in the local area, but full and honest disclosure requires you to note that these gains and losses in the local area are usually offset by changes elsewhere in the economy leading to no overall net change in employment at the state and national level.

To see how the transfer of income and employment works, consider the decision of whether an agency should approve a new shopping center in an area that is environmentally sensitive because it contains a wetland. The developer will always tout the direct jobs the new shopping store "creates." The stores will have to hire people as stock clerks, cashiers, etc. But, are these really new jobs to the country's economy? There are several reasons the answer to this question is no. First, where will the workers come from to work at the new stores? If there is full employment, the retail store workers will probably come from other existing jobs elsewhere. Now, think about what happens if the shopping center is not approved. Does it mean that all of the people who would have been employed at the grocery store will not have jobs for the proposed 30-year life of the stores? Probably not. Most of us have some kind of job to feed ourselves and pay rent.

To make the transfer even more obvious, think about what happens if the new shopping center is approved. Shoppers will come to this new store and spend money. But, is this really new money coming into the economy? It came from reduced spending at some other stores. For example, increasing the number of grocery stores by 30% does not mean people are going to eat four meals a day. The opening of a new shopping center simply rearranges existing spending: there is really no net gain in spending or employment in the country's economy. Unlike a new local shopping center, operation of new businesses that produce goods for sale or export outside the local regional economy (i.e. basic industries) will tend to have a more positive net effect on the local economy, since sales of the firm may bring new money into the local economy. Thus it is important for the policy analyst to carefully distinguish between expansion of basic or exporting sectors of the economy and expansion of businesses that cater primarily to local demand.

8.1 When Impacts May Also Be Economic Benefits

If the regional economy for a proposed project has substantial and persistent unemployment of labor, then the primary benefits of the project may include the income (salaries and wages) of otherwise unemployed labor. In the U.S., substantial and persistent unemployment is defined to exist in an area when unemployment in the previous year was 6 percent or more; and was at least 50 percent above the national average for three of the previous four years, or 75 percent above the national average for 2 of the previous 3 years, or 100 percent above the national average for 1 of the previous 2 years (U.S. Water Resources Council, 1983). Even in regions qualifying as having substantial and persistent unemployment, not more than half of the labor is assumed to be supplied by unemployed workers and can be counted as a benefit.

In the 1960s and 1970s the regions of the U.S. in which these conditions prevailed were the Appalachian Mountains, the Northern Lake states, and the Four Corners area of the Southwest. In these cases, secondary economic benefits measured by direct and indirect gains in local income would be counted as benefits, representing real economic gains to the national economy. Aside from these regions, few other areas would qualify in most developed countries. This is not to downplay the economic importance of regional economic impacts to the economic and political considerations. Much of the political motivation for development represents an attempt to capture regional gains, which in many cases are reflected in large increases in property values for commercial landowners or additional revenues for local business owners. From the viewpoint of the region in which a proposed project will be located, the residents affected are concerned about the gains and losses to themselves, not to other regions or the nation as a whole. But, when there is some federal taxpayer financing or federal permits involved, a policy analysis would be more complete if the national viewpoint was presented as well as the local.

Thus some policies that may appear beneficial to the local region are often essentially redistributing income from other regions, consumption sectors, and the nation. Whether such redistribution is desirable is a political decision typically beyond the scope of an objective policy analysis. Economists refer to such transfers of income as pecuniary impacts. This transfer also illustrates the zero-sum nature of chamber of commerce-like promotions in one town versus another: one's gain is often the other's loss.

Thus it is important for the policy analysis to state the local economic effects while noting consequences to other sectors or regions, so as to avoid becoming a policy advocate.

9. CONCLUSION

Almost every environmental impact statement discusses regional economic effects, and public debates about many environmental policies frequently include discussion of their local economic impacts. Input-Output models or multipliers derived from them are commonly used to quantify these local effects. This information is often quite helpful to decision makers and public officials in informing them as to whether local economic effects are or are not a real concern in choosing among policy alternatives. This information on local distributional equity helps to complement the economic efficiency analysis that focuses on overall or national economic analysis.

Increases in income in one county are rarely net economic gains but rather just transfers of economic activity from one region of the country to another. As such, changes in local county income cannot be compared to the overall project costs to arrive at meaningful measure of the economic desirability of a project if those costs include costs from outside the area (e.g., federal tax dollars). This mistake is frequently made since the multiplier analysis results in dollars of income.

The only figures that can be meaningfully compared are the local change in income or employment to the local share of the costs. This would yield a "local" benefit-cost analysis. But, as the discussion of externalities and national economic efficiency reveals, this very partial local analysis is likely to be quite misleading. Such a local benefit-cost analysis would ignore both benefits and costs accruing to people outside the local economy. Since these are very real benefits and costs, they should not be ignored. Thus, while it is often useful to display local benefits and costs separate from overall benefits and costs, it is important to keep in mind which is the more complete measure.

Chapter 11

Optimization and Linear Programming

In Chapter 3, we discussed optimization methods as a possible decision rule. Optimization methods typically have the form of maximizing or minimizing a goal, while assuring that a set of constraints is satisfied. That description makes this approach sound very complicated, and it can be. What may be surprising, once you start recognizing these problems, is how common they are. A tremendous number of decision questions can be viewed as constrained optimization problems. You may be surprised how many of your day-to-day choices -- how much time to spend working on something when you could be sleeping, how to allocate your paycheck, where to go on vacation -- fit into this framework.

Understanding the components of optimization problems and some of the basic principles of their solutions can by themselves be very helpful. For example, knowing that, in principle, adding a constraint can never improve the value of the objective function can make you think twice about the value of adding additional requirements to a proposed plan. Forcing someone to think about what, exactly, the objective is in a situation can help to clarify the issues at hand. Even if the problem is not explicitly mathematical, then, principles of the mathematical solution can help to understand the problem better.

Many environmental and resource policy decisions, too, are constrained optimization problems. Resources -- money available for pollution abatement, money for improving the conditions of national parks, personnel available to monitor compliance with the law -- are often limited (constrained), and they must be allocated among a number of competing goals. Should the U.S. Environmental Protection Agency (EPA) spend more of its budget on improving water quality, on improving air quality, or on regulating pesticide use? Which national parks should receive higher maintenance budgets? How much timber can be produced from the national forests without damaging ecological resources? All these cases are constrained optimization problems.

Chapter 3 discussed constrained optimization as a decision method -- that is, as a way of identifying the optimal solution to the problem being considered. Optimization methods will certainly do that, but they can serve many more purposes as well. They are very helpful for evaluating the effects of different scenarios, for instance. They can be used for sensitivity analysis, to determine if the optimal solution is sensitive to various values or assumptions in the problem. Sometimes they can be helpful just in an accounting sense: they can keep track of the effects of changing assumptions on a wide number of impacts. As a result, while constrained optimization can provide optimal solutions, it can also help to fill in the matrix of impacts for different alternatives without necessarily being used to decide among the alternatives.

This chapter will first review the general elements of a constrained optimization problem. These problems can be either non-linear or linear (in a way that will be described more thoroughly below); some aspects of these two forms differ sufficiently that they will be compared. Finally, we'll provide some examples of optimization.

1. THE ELEMENTS OF A CONSTRAINED OPTIMIZATION PROBLEM

All constrained optimization problems have three key elements: the objective function, the constraint set, and the activities (Stokey and Zeckhauser, 1979). Each of these will be described in turn below.

1.1 Activities

The activities in a constrained optimization problem refer to the choice variables. In other words, what are the activities that the decision-maker can control that determine the impacts? Identifying the activities is a critical part of the programming problem. For instance, consider the problem of allocating the EPA's budget among water programs, air programs, or pesticide programs. The activities, in this case, are the amount of money that goes to each program. In a land management case, such as faced by the U.S. Forest Service, the agency has to decide what management strategy to apply to each acre. Should a given site be used for intensive timber management, for a campsite, or for undeveloped recreation? The activities, then, are the choices of management practices for each acre. In a personnel allocation decision, the activities are the person-hours dedicated to each possible personnel use.

The activities are the backbone of the programming problem. The solution to the optimization problem is the allocation of these activities among the possibilities. Once the activities are allocated, the effects of this selection can be calculated. For instance, if the optimal allocation puts $1 million more of the budget into improving water quality, then the effects of that additional $1 million on the water quality program can be measured (and the effects of taking that $1 million out of the air and pesticide programs can also be measured). (As will be discussed, the effects of that additional $1 million are built into the optimization problem, through either the objective function or the constraints.)

Both the objective function and the constraints are functions, either directly or indirectly, of the activities. For instance, the EPA budget constraint – the limit on total funding that requires EPA to decide how to allocate its budget – is the sum of the allocations to the water, air, and pesticide programs, the three activities for this problem. In the Forest Service problem, the levels of timber, recreation, and wildlife that will result from the program, which are likely to be part of the objective function, depend on how many acres are dedicated to each of the possible management strategies. Thus, just defining what are the activities can be a useful exercise: these are the decisions that have to be made, and all other impacts will result from these choices.

1.2 Constraints

Constraints, as suggested by the name, are limits on the solution to the problem. For instance, the total budget that the U.S. EPA can allocate to its programs is a constraint. Identifying the total amount of forested land available for timber harvest is another example of a constraint. As these two examples show, the constraints on the equation can serve a number of different purposes, including providing definitions, requiring achievement of specified policies or physical limits, and giving an opportunity for sensitivity analysis.

Constraints can sometimes be usefully categorized as either "technical" or "policy" constraints. Technical constraints serve primarily the definition function: they describe the relationship between the activities and other variables of interest. For instance, a technical constraint would note that 1 acre of intensive forest management is associated with 10 tons of sediment;

another might note that that acre in intensive forest management would provide 15 thousand board feet (mbf) of timber. Additionally, technical constraints would identify what the EPA would do with money in each of its programs: for instance, each million dollars spent on water quality might reduce outflows of biological oxygen demand (BOD, a common pollutant that takes oxygen out of the water and thus damages fish habitat) by 1000 tons, while each million dollars spent on air quality might reduce emissions of sulfur dioxide by 10,000 tons.

In each of these examples, the technical constraints transform the activities – the acres assigned to intensive management, or the budget spent on each EPA program – to outputs or effects of interest, such as timber or sediment production or reductions in emissions. Constraints can also take these newly calculated variables and translate them into other variables of interest. For instance, the sediment measure can be converted via a technical constraint into effects of sediment on fish habitat or downstream hydroelectric production (because sediment, when it builds up behind a dam, reduces the efficiency of electricity production). Without further basis for comparison, you have no way of knowing whether reducing BOD by 1000 tons is more desirable than reducing sulfur dioxide by 10,000 tons: though both are measured in tons, you would need additional constraints to tell you which causes more damage to people, fisheries, or other variables of interest. Even though these new constraints are likely to be defined in terms of BOD and sulfur dioxide, both of those variables rely on the underlying activities, the amount spent on each program. Thus, as described above, the activities are at the root of all the constraints, either directly or indirectly.

Policy constraints also rely on activities, and they also can be built of new variables constructed from the activities. In contrast to technical variables, though, policy variables are not technical relationships; instead, they are limitations imposed on the problem by some external source. For instance, the total budget that EPA is given, to allocate among its programs, is a policy variable. Nothing in the laws of nature dictates that budget; it comes from decisions in Congress. Similarly, the Clean Water Act may put restrictions on sediment loads into streams, while the Endangered Species Act requires maintenance or improvement of population levels of endangered species. Policy variables are, in essence, human requirements imposed on the system rather than technical relationships and thus, at least in theory, can more easily be changed.

The distinction between technical constraints and policy constraints is much fuzzier than the above description suggests. Changing legal requirements can be politically difficult and can take long periods of time; even if, in theory, they are subject to the whims of policy, those whims of policy could functionally be permanent fixtures. And technical relationships can sometimes be changed, through changes in technology or practice: for instance, changing the way that timber is harvested, even in an intensive timber management regime, can affect the amount of sediment produced; more cost-effective pollution abatement equipment can increase the amount of pollution abated per million dollars of expenditure. While it can be useful to think about how inflexible any particular constraint is, you should not bother spending a lot of time deciding whether a constraint is technical or policy.

The constraints define the "feasible set" for the optimization problem: that is, they determine what combinations of the activities can achieve all the constraints. Consider, for instance, the EPA budget allocation problem, simplified (for graphical purposes) to the tradeoff between the air quality program and the water quality program, with a total budget of $10 million. The mathematical representation of this constraint is:

(11-1) Air Budget + Water Budget ≤ $10 million

Figure 11-1 shows a graph representing this constraint. With Air Budget on the X axis, and Water Budget on the Y axis, this constraint is graphed as a straight line representing all the combinations of the budgets that total to $10 million: splitting the money evenly between the programs, giving all to one and nothing to the other, etc. If this is the only constraint on the system, and if negative budget allocations are impossible, the triangle under this line but northwest of the origin is now the feasible set: no possible budget allocation outside this triangle achieves the constraint. If a further constraint is added – for instance, that a water quality mandate requires removal of 2 tons of BOD at a cost of $1 million/ton, then this constraint can be represented mathematically in two parts:

(11-2) Tons of BOD removed ≥ 2

(11-3) Water Budget = $1,000,000 * Tons of BOD removed

(notice that the second constraint implicitly makes the first constraint a function of the activity, Water Budget). When these two constraints are combined, they become

(11-4) Water Budget ≥ $2,000,000.

This constraint can be added to Figure 11-1 as a horizontal line at the level of $2,000,000 on the Y axis. Now the feasible set has shrunk: the area below the $2,000,000 line now includes points that do not satisfy all

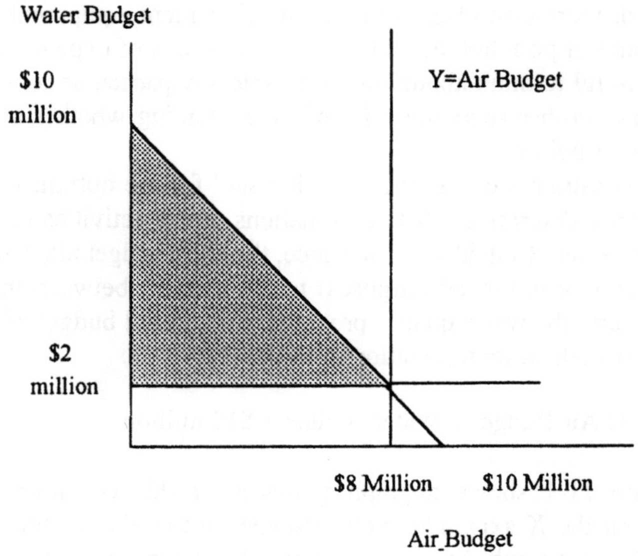

Figure 11-1 Optimal EPA Budget Allocation

the constraints. Only the area bounded by the $2,000,000 line, the Y axis, and the budget constraint is now feasible.

This example demonstrates a simple but very powerful concept known as LeChatalier's Principle: addition of a constraint can never improve your choice set and is likely to make it smaller. In other words, a constraint cannot make you better off; put yet another way, you are better off with more choices than with fewer. After all, if you have more choices, you don't have to choose the new choices, but you still have all the old choices available; you can't therefore be worse off, and you may well be better off, if you can increase your feasible set. This principle can be applied to a great many circumstances. Is more money better than less? Should students be required to take specific courses, or should they have greater choice? Are advances in technology good? In all these cases, LeChatalier's Principle will tell you that new options do not have to be accepted, but having them available provides additional, possibly better, opportunities.

(Mathematically, LeChatalier's Principle is always true. In terms of personal decision-making, there may be limitations on it. When you are already having trouble choosing between two courses, is the availability of a third option really likely to help, or will it only make the decision harder? More choices add costs to the decision-making process that are not explicitly part of this principle.)

It is quite possible to put so many constraints on a model that there is no feasible set. For instance, consider the budget problem above. Suppose that at least 15 tons of BOD were required to be removed from the system at a cost of $1,000,000 per ton. Since this constraint requires $15,000,000, but only $10,000,000 is available, there is no allocation of money that satisfies these constraints. Constraints, while an important (and often unavoidable) part of an optimization problem, can also prevent a solution. Restraint should be used in putting each additional constraint onto a model.

Not all constraints limit the size of the feasible set. For instance, suppose that another constraint in the EPA budget problem required that the water budget exceed $1 million. Obviously this constraint will not affect the solution to the problem, since another constraint requires at least $2 million in the water budget. If a constraint does not affect the solution, it is said *not to bind*; in contrast, if a constraint does change the solution to the problem, it is said to be *binding*.

The constraints thus both define and limit the choices available for the problem at hand. In their role of providing definitions, they identify key relationships and keep track of impacts. As limitations due to policy decisions, they identify the impacts of those choices. As will be seen below, adding, deleting, and changing constraints can provide important insights into the model and the results from it.

1.3 The Objective Function

As just discussed, the constraints define the feasible set. If there is more than one set of activities in the feasible set, how should the choice be made among those points? The final part of the problem, the objective function, specifies the criterion by which the choice is made. Is the goal to maximize net benefits to society? To minimize adverse effects on human health and the environment? To maximize profits or minimize costs? In other words, what is it that you really want to accomplish?

The objective function is a function, either directly or indirectly, of the activities. It can be as simple as maximizing or minimizing one activity (e.g., maximize the air budget) or one good related to an activity (maximize timber production), or it can be a more complicated function of multiple activities and goods. For instance, in the EPA budget problem, the objective might be to minimize harm to human health and to the environment. "Human health" and "the environment" would then have to be defined: for instance, human health could be deaths due to lung cancer and water-borne contaminants; "the environment" could be damages to plants and aquatic systems. All those effects would then have to be measured in terms of the budgets to air and water quality activities: for instance, each ton of BOD removed (at $1 million/ton) might be associated with avoiding 1 death due to water-borne contaminants as well as protecting 15 trout. (Those effects could end up in the constraint set, or they could be substituted directly into the objective function.) In all these cases, the objective function provides an explicit statement of goals. It therefore provides the valuable function of forcing a policy-maker to make priorities not only clear, but mathematically formulated.

The objective function can originate from a variety of places. Legislative mandates may specify the function, either explicitly or implicitly: for instance, the Clean Air Act's requirement for air quality standards could be phrased as minimizing effects on human health. In the private sector, profit maximization and cost minimization are typically considered reasonable approximations of how businesses operate. In other cases, an agency or policy-maker may want to convene staff to discuss the objectives and how to assign weights to different parts of the problem. Prices might be appropriate weights to use in some cases, since (as discussed in Chapter 3) they provide a way of converting apples and oranges into comparable monetary units. In other cases, non-market valuation techniques (as discussed in Chapter 9) may be necessary to provide weights for some of the relevant goods. Having an explicit discussion about objectives can be a difficult but ultimately quite useful activity for a decision-maker.

Once the objective function is specified, its role is to identify the levels of all activities that satisfy the constraints and also best meets the specified goals. For instance, in the EPA budget problem, suppose that the goal is to maximize the air budget subject to the total budget constraint (Air Budget + Water Budget ≤ $10 million) and to the total BOD constraint described above (Tons of BOD removed ≥ 2, which was converted to Water Budget ≥ $2,000,000). The typical way that this set of equations is presented is:

(11-5) $\text{Maximize}_{\text{Air Budget, Water Budget}}$ Air Budget

subject to:

(11-6) Air Budget + Water Budget ≤ $10 million

(11-7) Water Budget $\geq$ \$2,000,000

The objective function is presented first, with an explicit statement of what the goal is (usually to maximize or minimize something). The activities are listed, so that it is clear what are the variables that can be manipulated. The phrase "subject to" refers to the choice of activity levels being subject to their being in the constraint set.

The feasible set for this problem has already been described and is graphed in Figure 11-1. The objective function will now determine the combination of activities in the feasible set that best meets the objectives set out. As a function of the activities, it too can be graphed on this figure, as an equation of the form Y = Air Budget, where Y is the value that the objective function achieves. Initially, that value is not known. The equation can be graphed as a vertical line, and it is clear that increasing Air Budget involves moving the line to the right. The furthest right line that still touches the feasible set solves this programming problem. That solution is shown as:

Air Budget = \$8,000,000

Water Budget = \$2,000,000

It is easy to verify that both constraints are satisfied. It is also easy to verify that Air Budget is at the maximum possible given these constraints. In general, the objective function becomes a line with a (possibly multi-dimensional) slope and an unknown intercept; moving the objective function either in (to minimize) or out (to maximize) over the feasible set until it just barely touches the edge of the feasible set provides the solution to the optimization problem.

Just as the constraint set frequently has flexibility in it, the objective function can be varied to perform sensitivity analysis. Different weights can be assigned to the elements in the objective function, or the overall form of the objective function can be changed completely. It is often true that the objective function is subject to some debate; perhaps people have different goals, or perhaps the weights assigned to the elements in the objective function are uncertain. Regardless, changing the objective function permits experimentation with the sensitivity of the answer to changes in this function. If the same answer occurs even with different forms of the objective function, then that set of activities is likely to be a good choice; if the answer is highly sensitive to this choice, refining the objective function could be important.

Changing the objective function can also serve as a way to formulate alternatives and to examine their effects at the same time. For instance, the objective function in the EPA budget problem could be to minimize harm to human health, or it could be to minimize adverse environmental impacts, or it could be to maximize net benefits to society. All these objectives, as described above, would have to be defined in terms of the activities, and the same set of constraints could be used with these three different objective functions. The resulting sets of activities could be the basis for different alternatives.

1.4 Shadow Values

Although it is probably not strictly correct to consider shadow values as elements of optimization problems (they are, more correctly, a result of them), they can provide very useful information about the problem. A shadow value is the marginal effect of the constraint on the objective function: that is, if the constraining factor is changed by one unit, the shadow value will measure the effect of that change on the objective function. If a constraint is not binding, the shadow value is zero: there is no cost to the constraint, because it does not affect the solution or the level of the objective function. In contrast, if the constraint is binding, then the shadow value does affect the objective function and is positive.

Shadow values may be calculated automatically by optimization packages. They can help determine which constraints are actually affecting the solution to the problem, and which do not influence it. (In complicated models, it is often not possible to tell by inspection which constraints bind and which do not.) If a constraint binds and has a large shadow value, it may be worth asking how inflexible the constraint really is, since a large improvement in the value of the objective function may come from a small change in the level of the constraint.

Additionally, the shadow value may implicitly provide a dollar value for a non-marketed good. Shadow values are measured in the units in which the objective function is measured, divided by the units in which the constraint is measured. For instance, if the objective function is measured in dollars, and the constraint limits the number of acres, the shadow value is measured in dollars per acre. Suppose, as in the forest management example below, that the objective function is measured in dollars, and there is a constraint relating to a non-marketed good (such as a restriction on sediment production or a requirement to maintain wildlife habitat). The shadow value of the constraint indicates how much the marginal ton of sediment or the marginal acre of habitat changes the objective function -- in other words, the value associated with the last unit of the constraint. It then becomes possible to discuss with a policy-maker whether restricting sediment is worth, for instance, \$50/ton, or whether an acre of land dedicated to habitat is worth \$1000 in foregone net benefits. Implicitly the optimization model determines the cost of the constraints; policy-makers can then discuss whether the benefits of the constraint are likely to outweigh those costs.

In sum, these three elements – activities, constraints, and objective function – form the basis of any optimization problem. Both the constraints and the objective function need to be defined, either explicitly or implicitly, in terms of the activities. The constraints determine the feasible set – the range of values possible for the activities, given technical, legal, or institutional requirements. The objective function determines the best set of activities in the feasible set for achieving the specified goal. Finally, the shadow values of the constraints, which result from these three elements, identify which constraints influence the solution to the optimization problem, and which do not affect the solution. The following section gives a more detailed example of this optimization exercise.

2. EXAMPLE: NATIONAL FOREST MANAGEMENT

The United States Forest Service (USFS) manages about 190 million acres of land in the U.S. The guiding principles for management of these lands are found (in part) in the Multiple Use and Sustained Yield Act of 1960 (MUSY); it specifies that the national forests "are established and shall be administered for outdoor recreation, range, timber, watershed, and wildlife and fish purposes." (Sec. 1) Multiple use is defined as (Sec.4(a))

> the management of all the various renewable surface resources of the national forests so that they are utilized in the combination that will best meet the needs of the American people; making the most judicious use of the land for some or all of these resources or related services over areas large enough to provide sufficient latitude for periodic adjustments in use to conform to changing needs and conditions; that some land will be used for less than all of the resources; and harmonious and coordinated management of the various resources, each with the other, without impairment of the productivity of the land, with consideration being given to the relative values of the various resources, and not necessarily to the combination of uses that will give the greatest dollar return or the greatest unit output.

The above definition is elegantly written, even by legal standards. To make it practical, though, the above statement can also be translated into optimization language. In this version, multiple use is intended to "best meet the needs of the American people" through management for "outdoor recreation, range, timber, watershed, and wildlife and fish purposes" "without impairment of the productivity of the land." With that reinterpretation (and gross oversimplification), the activities which the USFS is supposed to control are "outdoor recreation, range, timber, watershed, and wildlife and fish purposes;" the constraint is that the "productivity of the land" cannot be impaired; and the objective function is to "best meet the needs of the American people," "with consideration being given to the relative values of the various resources, and not necessarily to the combination of uses that will give the greatest dollar return or the greatest unit output."

It still sounds too poetic to be mathematically solvable. It needs to be converted into formulas and algebraic notation. The USFS in the 1980s used an optimization model called FORPLAN to design management alternatives and to identify preferred alternatives for the national forests. The following example is s seriously oversimplified version of the kind of problem for which FORPLAN might be used.

Suppose there are 2 different land types on the national forest: forest and range. The different management practices that can be applied to these land types are: intensive forest management (IFM), intensive range management (IRM), recreation emphasis on either forest (RFM) or range (RRM), wilderness management on either forest (WFM) or range (WRM), and campsite, which requires a mix of 10 acres of range and 15 acres of forest. These constitute the activities: the fundamental choice that the USFS faces is what management activity to apply to each acre. In turn, the choice of management activity for each acre will determine the outputs from the forest, including grazing, recreational opportunities, wildlife habitat, and sediment runoff.

Table 11-1 determines the levels of outputs associated with each acre of each type of activity (except for campsites, each of which requires 10 acres of range and 15 acres of forest). Thus, for instance, an acre of forest dedicated to intensive forest management (IFM) will produce 20 thousand board feet (MBF) of timber, 0.2 animal unit-months (AUMs) of grazing, 0.5 tons of sediment, 0.3 acre-feet of water, and 1 recreation visitor-day (RVD), at a cost of $200/acre. This information can be converted into both technical constraints and policy constraints. For instance, suppose there are a total of 10,000 acres of forest and 15,000 acres of range available, and let IFM, IRM, etc. represent the number of acres allocated to each of the above activities. The constraints on total acres of forest and range thus respectively become:

(11-8) IFM + RFM + WFM + 10*Campsite = 10,000 Acres

(11-9) IRM + RRM + WRM + 15*Campsite = 15,000 Acres

Table 11-1. Outputs associated with land management practices

	IFM	IRM	RFM	RRM	WFM	WRM	Campsite	$ per Unit
Timber (MBF)	20	0	1	0.2	0	0	0	$200
Grazing (AUM)	0.2	2	0	0	0	0.05	0	$20
Sediment (Tons)	0.5	0.3	0.05	0.05	0.01	0.01	0.25	
Water (Acre-Feet)	0.3	0.3	1	0.5	1.5	1.5	3	$50
Recreation (RVD)	1	1	100	60	10	15	2000	$30
Cost	$200	$100	$50	$100	$10	$10	$2000	

Total timber production can be found by summing the timber produced per acre times the number of acres in each use, or

(11-10) Total Timber = 20*IFM + 0*IRM + 1*RFM + 0.2*RRM + 0*WFM + 0*WRM + 0*Campsite

This equation demonstrates how an output (in this case) can be constructed from the activities. This output, timber, can then be included in the objective function, if timber is one of the factors on which management of these lands depends. Alternatively, if law, regulation, or other factors suggest a target (either maximum or minimum) on total timber production, this equation could become another constraint. The choice of whether timber should be in the objective function or the constraint set depends on whether it is considered part of the objectives for the land, or whether a specified level of production is to be achieved. It is also legitimate to include timber both in the objective function and in the constraint set; if it is included in both places, the solution to the problem will determine if the constraint binds (i.e., the constraint determines the level of timber production), or whether the constraint does not bind (the constraint is satisfied regardless of its inclusion in the programming problem).

Similar equations can be developed for sediment, recreation, grazing, water production, and the costs of providing these activities. They too can be included either in the objective function or among the constraints. In either case, the definitional equation (like Equation 11-10) will be part of the technical constraints of the problem: they convert the activities into the goods for which,, fundamentally, the USFS is managing. The choices of activities are means to ends; the ends are these goods.

As discussed above, the objective function comes from the preferences of the decision-makers. Also as discussed above, choices can be made about whether a good should be included in the objective function or in the constraint set. Finally, the objective function can be used to understand the nature of the problem and to develop alternatives. Suppose, for instance, that the USFS wants to know what the maximum amount of timber production from this forest could be if the only environmental constraint were a limit on 500 tons of sediment. The objective function for this problem would then become:

(11-12) Maximize Timber = 20*IFM + 0*IRM + 1*RFM + 0.2*RRM + 0*WFM + 0*WRM + 0*Campsite,

and the constraint set would be:

(11-13) IFM + RFM + WFM + 10*Campsite = 10,000

(11-14) IRM + RRM + WRM + 15*Campsite = 15,000

(11-15) Sediment = 0.5*IFM + 0.3*IRM + 0.05*RFM + 0.05*RRM + 0.01*WFM + 0.01*WRM + 0.25*Campsite ≤ 500

The solution to this problem has the following allocation of management practices.

Table 11-2. Land allocation from maximizing timber with 500-ton sediment constraint

IFM	IRM	RFM	RRM	WFM	WRM	Campsite
510.2	0	0	0	0.1	765.46	948.97

The optimization model has not only provided the land allocation that maximizes the amount of timber possible, but it also keeps track of all the other effects -- in this case, the following set of outputs.

Table 11-3. Outputs from maximizing timber with 500-ton sediment constraint

Timber (MBF)	Grazing (AUM)	Sediment (Ton)	Water (Acre-Feet)	Recreation (RVD)	Cost	Revenue
10,204	140.31	500	4,148.3	1,909,932	$2,007,636	$59,549,005

All three constraints are shown to be binding, with shadow values:

Table 11-4. Shadow values for maximizing timber with 500-ton sediment constraint

Forest Total	Range Total	Sediment Total
-0.0482	-0.0482	40.816

In this case, probably because of the sediment constraint, increasing forest or range acreage actually *reduces* the amount of timber available, with each acre costing production of 0.05 MBF. In contrast, lifting the sediment constraint by one unit would increase timber production by 40.8 MBF. It is thus clear from examination of the shadow values that the sediment constraint has a significant effect on timber production from these lands.

It is also possible to identify the power of the constraints through changing them. For instance, if the sediment constraint is lifted to 1000, the allocations and impacts change:

Table 11-5. Land allocation from maximizing timber with 1000-ton sediment constraint

IFM	IRM	RFM	RRM	WFM	WRM	Campsite
1530.61	0	0	0	0.05	2296.00	846.93

Table 11-6. Outputs from maximizing timber with a 1000-ton sediment constraint

Timber (MBF)	Grazing (AUM)	Sediment (Ton)	Water (Acre-feet)	Recreation (RVD)	Cost	Revenue
30,612	420.92	1000	6444	1,729,839	$2,022,950	$58,348,227

while the per-unit shadow values are the same as above.

About three times as much timber is produced, since about three times as much land is put into intensive forest management, with the relaxed sediment constraint. Thus, the sediment constraint significantly affects timber production in this area. If maximizing timber production is the goal for these lands, the USFS should consider seriously how much sediment it is willing to accept from the forest.

Alternatively, the USFS may be interested in promoting more recreational use of its lands. If it sought to maximize recreational use (or just to identify the maximum recreational use) with the 500-ton sediment constraint in place, the solution would be

Table 11-7. Land allocation from maximizing recreation with 500-ton sediment constraint

IFM	IRM	RFM	RRM	WFM	WRM	Campsite
0	0	0	0	0	0	1000

with the following set of outputs:

Table 11-8. Outputs from maximizing recreation with a 500-ton sediment constraint

Timber (MBF)	Grazing (AUM)	Sediment (Ton)	Water (Acre-feet)	Recreation (RVD)	Cost	Revenue
0	0	250	3000	2,000,000	$2,000,000	$60,150,000

and shadow values:

Table 11-9. Shadow values from maximizing recreation subject to a 500-ton sediment constraint

Forest Total	Range Total	Sediment Total
110	60	0

All the land is organized into campsites, and no production of other resources occurs except incidentally to recreation production. Now, additional land would be useful (especially forest land) for producing more recreation, but the sediment constraint is not binding.

These two alternatives show how changing the objective function alters the chosen allocation. It also demonstrates how optimization models can be used to understand better the nature of the problem that the agency is trying to solve: there are now upper bounds placed on both recreation and timber production, given a 500-ton sediment constraint. (Note, though, that the sediment constraint does not bind in this scenario: when recreational use is maximized, sediment production is only 250 tons.)

Of course, the USFS's legal mandate is not for timber production; it is for multiple uses. Thus, it actually wants to manage for other outputs as well. It is tempting to say that the objective function should be Timber + Grazing + Water + Recreation, but such addition is almost meaningless: all these outputs are measured in units that are not comparable. A board-foot of timber and an acre-foot of water cannot be added directly. Thus, it would be useful to have weights for these outputs that would convert them to the same units, where those units would represent their relative value to society. As discussed in other chapters, prices of goods are one measure of relative value. Suppose, as shown in the table, that timber is worth $200/MBF, grazing is $20/AUM, water is $50/Acre-Foot, and recreation is worth $30/RVD. Then the objective function might be:

(11-16) Maximize 200*Timber + 20*Grazing + 50*Water + 30*Recreation,

and this total is now measured completely in dollars. With this objective function and the 500-ton limit on sediment, the resulting allocation and outputs are:

Table 11-10. Land allocation from maximizing revenue subject to a 500-ton sediment constraint

IFM	IRM	RFM	RRM	WFM	WRM	Campsite
454.55	0	0	681.82	0	0	954.55

Table 11-11. Outputs from maximizing revenue subject to a 500-ton sediment constraint

Timber (MBF)	Grazing (AUM)	Sediment (Ton)	Water (Acre-feet)	Recreation (RVD)	Cost	Revenue
9227.27	125	500	3340.9	1,950,455	2,068,182	60,528,636

and shadow values measured in dollars per acre or ton.

Table 11-12 Shadow values from maximizing revenue subject to a 500-ton sediment constraint

Forest Total	Range Total	Sediment Total
3291.73	1790.27	1514.54

The allocation has changed from the timber or recreation maximization scenarios: less timber and recreation are produced (of course, because those scenarios should produce the maximum timber or recreation possible). It is notable, though, that timber and recreation production in this scenario are within 10 percent of those maxima. An alternative in between maximizing timber and recreation produces most of the timber and recreation benefits of managing only for one or the other output. Thus, multiple use management appears to be both possible and advantageous for this area.

In all these versions of the model, sediment control has been achieved by making it a constraint rather than including it in the objective function. Another way of including sediment into the problem is by assigning it a value in the objective function. For instance, suppose that research has determined that sediment imposes costs of $1800 per ton, and the objective function now becomes:

(11-17) Maximize 200*Timber + 20*Grazing + 50*Water + 30*Recreation – 1800*Sediment with the only constraints being the total areas of forest and range. The resulting solution puts all the land into campsites, with the same outputs as the recreation maximization alternative. On the other hand, if sediment's cost is $1200/ton, the resulting solution is:

Table 11-13. Land allocation from maximizing revenue with sediment cost = $1200/ton

IFM	IRM	RFM	RRM	WFM	WRM	Campsite
10,000	0	0	15,000	0	0	0

Table 11-14. Outputs from maximizing revenue with sediment cost = $1200/ton

Timber (MBF)	Grazing (AUMs)	Sediment (Tons)	Water (Acre-Feet)	Recreation (RVDs)	Cost	Revenue
203,000	2750	5,750	10,500	910,000	$3,500,000	$61,580,000

with shadow values measured in dollars per acre.

Table 11-15. Shadow values from maximizing revenue with sediment cost = $1200/ton

Forest Total	Range Total
3449	1806

Timber becomes valuable enough relative to sediment to assign all forests to intensive timber management, and range management becomes focused on grazing production in spite of the sediment it produces. There is no longer a constraint associated with sediment; as a result, there is no shadow value associated with it. Because the land allocations and outputs are so radically different when the cost associated with sediment changes, it is clear that the value used for sediment should be identified with as much precision as possible before it is included in the objective function.

This example is intended to show that an optimization model is a very versatile tool. While it can solve specific problems with well known and well formulated parameters, it can also be used to explore more thoroughly the implications of the problem at hand. As discussed above, once the basic model is developed, changing any of the parameters of the model might change the results; thus, it is easy to identify how sensitive the problem is to the chosen parameters. It can also be used to explore the maximum or minimum possible outputs associated with the model, and to develop alternatives. Additionally, once the optimization problem is solved for the activities, the model immediately provides the values of everything defined in terms of those activities. Thus, an optimization model can also be very helpful in filling in the matrix of impacts of an alternative. Optimization models also allow answering "what if" questions about a management action prior to undertaking the management actions. Finally, if an alternative arises from a specific objective function and set of constraints, that alternative can then be linked clearly to a specific objective and set of constraints: it is obvious what the alternative is intended to do (and not to do). Optimization models can be complicated and difficult to build, but they can serve a number of purposes once constructed.

3. BUILDING AN OPTIMIZATION MODEL

The example just provided included a great deal of information about the effects of a specified management practice on a number of different outputs. Where, you may ask, did all this information come from? (In fact, you *should* ask this question.) One of the challenges in building an optimization model is that it may require large amounts of information. As this example shows, the information can take many forms: it may be on physical or biological effects, and it may also encompass economic or social variables. The information in the model can come from a number of places: previous studies, mathematical models, original data collection and analysis, and expert opinion are some of the many places that these data might have originated. Even regression estimates (see Chapter 6) can be used to define relationships among activities.

Probably the most important factor to understand about the data you are using in your model is the level of certainty that you associate with the data. In the model above, the effects of each management practice on, e.g., sedimentation are presented as certain, and the answers are found to multiple decimal places. In fact, though, the results may be very sensitive to the sedimentation values presented in the model. How radically would land allocations change if the sediment values were changed? Comparing the results of the 500-ton sediment constraint to the 1000-ton limit shows that sediment values do indeed change the optimal land allocations quite substantially. (Changing the constraint to 1000 tons plays the same role as dividing all the sediment per acre values in half while keeping the 500-ton constraint.) Thus, the precision of these values is important to the outcome of the analysis. Because these numbers do matter, it may be important to put more research effort into improving the certainty associated with those values. Similarly, changing the cost associated with sediment from $1800 to $1200 per ton changes the land allocation substantially. Thus, the relative values in the objective function also matter for the final results.

If the data in the model have a great deal of uncertainty associated with them, and if the results are highly sensitive to these uncertain values, you may want to reconsider using an optimization model. The precision that comes with the solutions from the model belies the fact that the data underlying the model have no guarantees associated with them. It may be more honest to develop alternatives based on other criteria than maximizing or minimizing different objective functions.

At the same time, the model may be useful, even with uncertainties, in its accounting role -- that is, in its ability to keep track of the impacts of different alternatives on the criteria of interest. The model can be set up with extreme values for, e.g., sediment load from a particular practice, and any land allocation entered into the model will then have associated with it a minimum and maximum expected sediment load. The model may no longer be an optimization model, but it can nevertheless assist in understanding the effects of alternatives.

4. LINEAR VERSUS NONLINEAR PROGRAMMING

All the concepts presented above apply to any optimization problem. For primarily computational reasons (and some theoretical reasons), it is nevertheless useful to divide optimization problems into two categories: linear programming and nonlinear programming. "Linear" here has a different meaning than when used for linear regression. In the case of linear regression, the problem has to be linear in the coefficients, but they can be nonlinear in the variables -- for instance, a regression can have independent variables for both Income and Income^2 . In contrast, in linear programming, all the equations -- the objective function and the constraints -- have to be linear in the activities. For instance, in the National Forest problem presented above, all equations are weighted sums of the activities, with no nonlinear transformations of the activities. Nonlinear programming allows the objective function, the constraints, or both to have the activities entered in nonlinear form.

Computer packages can solve linear programming problems much more easily than nonlinear problems. Thus, linear programming problems can be much larger than nonlinear problems and still be solved quickly and easily. Nonlinear programming problems may have many (local) solutions, and it may be difficult computationally to find the best solution. For a variety of reasons, though, it may not always be desirable to formulate a problem to be linear. The following discussion will highlight some of the important issues associated with their differences.

Linear programming problem requires that relationships between activities and other goods be linear. For instance, as described above, each acre put into intensive forest management produced 20 MBF of timber, 0.5

tons of sediment, 1 recreation visitor-day (RVD), etc. Under the linearity assumption, 1000 acres in intensive forest management would thus produce 20,000 MBF of timber, 500 tons of sediment, and 1000 RVDs. It may be, though, that these relationships are not in fact linear. One thousand acres of timber harvested, for instance, may lead to fewer than 1000 RVDs, as people dislike large harvested areas more than small harvested areas. Suppose that the true relationship between visitors and acres in intensive forest management is RVD = $(IFM)^{0.5}$. Then one acre in intensive forest management yields 1 RVD, as shown in the table, but 1000 acres in intensive forest management yields 31.6 RVDs, obviously much less than 1000. The assumption of a linear relationship may not be a bad approximation when numbers change only over a small range, but they may be quite wrong over a larger range (such as the difference between 1 and 1000 acres).

Another effect of the linearity assumption is that the solution is always found either at a corner of the feasible set (where two constraints intersect), or along one of the sides of the feasible set. Because all the elements of the constraint set are linear, the constraint set is a polygon. Because the objective function is linear as well, it will, by moving up or down as much as possible, touch the constraint set either at one of the corners of the polygon or along one of its sides. (The latter happens if the objective function has the same slope as one of the constraints. In this case, the solution is any point along that side of the polygon; any point on that line segment will produce the same value of the objective function.) The effect of these linearities is that the solution to the linear programming problem can, with small changes in the objective function, change dramatically. This effect can be seen through examination of Figure 11-1. There, the objective function hits its maximum at the intersection of the two constraints. This solution would still hold if the objective function were changed in any way that put slightly more weight on the Air Budget than the Water Budget -- for instance, if the objective function were to maximize 1.1(Air Budget) + Water Budget. If the objective function were, instead, to maximize Air Budget + Water Budget, the solution would be indeterminate: any combination of the Air and Water Budgets that achieved a total of $10 million, with at least $2 million for Water, would achieve the maximum. If the objective function were changed to put only slightly more weight on the water budget -- for instance, it was to maximize Air Budget + 1.1(Water Budget) -- the solution would now be at the top corner, with all the budget spent on water. The possibility that the solution to a linear programming problem might jump from one corner to another shows the sensitivity of these problems to the assumptions embedded in them.

Nonlinear programming problems are more likely to respond to small changes in the problem with small changes in the solution set. Suppose, for example, that each additional unit of timber harvested from an area reduced recreational use, and that relationship is estimated as:

(11-18) $RVD = 100 - Timber^2$

In other words, if there is no recreation in an area, 10 MBF of timber can be produced; if there is no timber harvesting, 100 RVDs can be produced; and combinations in between can be found from that equation (a production possibilities frontier, or PPF, which shows the maximum amounts of goods that can be produced from the same inputs -- in this case, the same piece of land). Figure 11-2 shows this PPF. If the value of timber is $200/MBF, and the value of a recreation visitor-day is $30, then the problem:

(11-19) Maximize 200*Timber + 30* RVDs

subject to:

(11-20) $RVD = 100 - Timber^2$

produces the solution:

Table 11-16. Solution to nonlinear programming problem with timber price = $200/mbf

Timber (MBF)	Recreation (RVD)	Objective Function
3.33	90	$3,366.67

If the value of timber rises to $210, the solution becomes:

Table 11-17. Solution to nonlinear programming problem with timber price = $210/mbf

Timber (MBF)	Recreation (RVD)	Objective Function
3.5	87.75	$3,367.50

In this nonlinear case, a small change in the weights in the objective function led to a small change in the allocation between timber and recreation, and a small change in the objective function. In the linear case -- for instance, if the constraint were RVD = 100 – 10*Timber -- the solution might not change at all, or it might shift suddenly from only recreation to only timber.

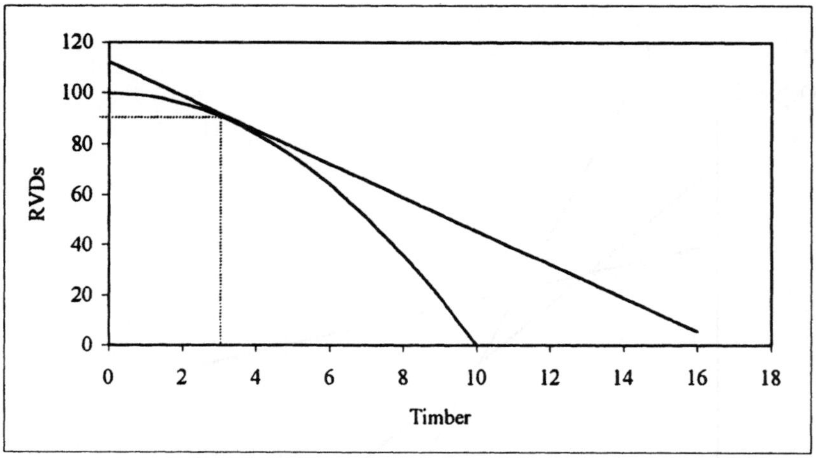

Figure 11-2 Optimization with a Nonlinear Production Possibilities Frontier

This characteristic of nonlinear programs -- the fact that they may change incrementally in the face of small changes -- is often a more accurate reflection of reality than the extreme choices of linear models. If the linear model is complex, so that there are many constraints, it may start to mimic a nonlinear problem, though, by the addition of more corners to the polygon. Figure 11-3 shows how multiple constraints can make a polygon resemble a curved surface. While small changes in the objective function can still make the solutions in the linear problem jump from one corner to another, the corners are now closer together. In complicated cases like this, the computational desirability of a linear programming problem may outweigh the possibly increased accuracy of a nonlinear problem. In contrast, simple nonlinear problems, such as the timber vs. recreation choice above, can be solved easily, even by hand; in these cases, keeping the nonlinearities of the problem intact may lead to a more accurate representation of the system.

The computational complexity of nonlinear models and the possible unrealistic nature of linear models are often the major determinants of the choice of a linear vs. a nonlinear model. Other factors also influence this choice, of course. Some problems, due to their nature, are inherently linear programming problems. For instance, Taylor, Adams, and Miller (1992) examined the choice of crops and farming activities on different types of land in the Willamette Valley in the face of different policies to reduce nitrate runoff. Each acre of a specified land type in a specific crop

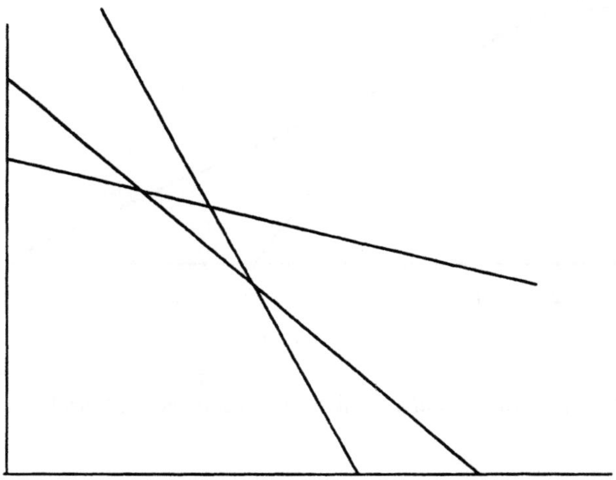

Figure 11-3 Series of Linear Constraints Creating a Non-Linear Effect

management scheme can easily be expected to produce the same profits; total profits are the sum of the choice of crop on each acre of each type of land; and runoff from each acre can be expected to be a constant. For this case, linear relationships are inherent in the problem, and a linear program is the appropriate tool. In contrast, Helfand and House (1995) wanted to examine how farmers would change their applications of inputs (water and fertilizer) in response to various policies to reduce nitrate runoff. Because the yield of the crop responded with diminishing marginal returns to increases in water and fertilizer applications, the problem was a nonlinear one, and they used nonlinear programming to determine the results. (A linear program would have implied that doubling fertilizer and water applications always doubled yield, a highly unrealistic assumption.) Linear programming also implies constant returns to scale, among other restrictive assumptions that may be more realistic in some settings than others. Thus, some problems are inherently linear, and others deserve to have their nonlinearities preserved.

As computational costs drop with the advent of more sophisticated computers and programming languages, it becomes easier to say that the nature of the problem should dictate the choice of method, rather than the computational ease. For simple problems, this condition is certainly true. If a model is large enough, though, that solving it via linear programming appears advantageous, it is important to recognize the effects of the linearity assumptions. Sometimes, as in the case of the Taylor, Adams, and Miller study, the linear approach will be appropriate for the problem; in other cases, it may be a significant limitation.

5. USES OF OPTIMIZATION

The discussion so far has described and alluded to many of the functions that an optimization model can serve. Clearly, the simplest role is that it can find the levels of activities that both satisfy a set of constraints and lead to a maximum or minimum value of the objective function. Additionally, as the discussion of shadow values has shown, the model can identify what constraints influence that allocation, and how costly they are to the objective function.

This latter function suggests (and the previous example shows) that an optimization problem does not have to be used to find the one and only solution. Instead, the model can be changed -- by adjusting either the levels of the constraints, the parameters in the model, the form of the objective function, or the weights in the objective function -- to understand better the ramifications of any one solution. All these possible manipulations are forms of sensitivity analysis, and they can be used to determine how sensitive a solution is to these factors. If the solution does not change, or does not change substantially, in response to these manipulations, then the result is said to be *robust*. Even if there are uncertainties about the parameters, they do not influence the result; further research into these parameters is not warranted. On the other hand, if the solution is highly sensitive to these manipulations, then developing more certainty about the parameters or functions could be very important.

Even the effort to formulate a policy question as an optimization model can be a valuable function of this method. Identifying the activities that the policy-maker can control (and, as well, the ones that she cannot control), identifying both the technical and the policy constraints that she faces, and forcing her to identify her objectives will make the entire problem clearer both to you as the analyst and to her. Are the constraints in fact immovable, or can they be bent? What value does she place on various activities or other outputs? Even if she cannot explicitly form her objective function for you, you may be able to show her the consequences of different objective functions. In the national forest management example, the optimization model found the maximum timber and recreation outputs possible with the sediment constraint in place, and the tradeoffs involved in providing one or the other (the marginal rate of transformation, in economic jargon).

Finally, but far from trivially, the optimization model can serve as an accounting tool. As the national forest example showed, having all the outputs measured in terms of the levels of the activities, and then solving for the activities, produces values for all the outputs as well. The model can thus be a very valuable method for filling in the matrix of impacts of different alternatives. If an alternative is viewed as a different run of the model (through a changed objective function or constraint set), then running the model can identify many (if not all) of the impacts of that alternative.

Nevertheless, as discussed above, optimization models may require tremendous amounts of data, and they may be only as good as the data used in them. If your model is built only on very rough guesstimates, the precision of your answer may be a pretense. Sensitivity analysis is then an important requirement. It may even be more appropriate not to use a formal optimization model, but instead to rely on other methods to identify solutions. The effort put into formulating the optimization problem, though, is not wasted. Just identifying the activities, the constraints, and the goals, even if not with precise numerical values, will help you as you tackle the problem with other tools.

Chapter 12

Risk Analysis

1. WHAT IS RISK ANALYSIS?

Almost every action in life carries with it the possibility of adverse consequences. Driving a car to work carries with it a chance of injury or even death. Yet millions of people voluntarily accept this possible outcome and drive to work everyday because the trade-off of convenience for risk of injury or death is worth it to them. Other times, the risks we face are not chosen by us, but imposed on us via decisions of firms in the market place or public agencies. Risk analysis helps decision-makers understand that the effects or outcomes of many policy actions are not known with certainty. There is sometimes a wide range of possible future outcomes, each with its own likelihood of occurring. For example, constructing a new landfill may result in a small chance of leakage of pollutants into the groundwater. Further, there is a small chance that such pollution in the groundwater may eventually end up in a town's drinking water supplies. If this were to occur, a small percentage of the population might become ill from this exposure. How is an analyst to determine if building this new landfill is worth the potential drawbacks? This chapter will outline the basic approach to incorporating these types of probabilistic outcomes into a policy analysis.

1.1 What is Risk?

1.1.1 Risk

In order to define risk analysis, we need to define what we mean by risk. Risk is generally defined as an outcome we are unsure of, but we do know the alternative outcomes or states of the world, *and* we can estimate the probabilities of these different outcomes (Zeckhauser and Viscusi, 1990). To quantify risk, first we need to know the possible outcomes and then the likelihood of each occurring. A simple example is a coin toss. We know the outcomes (heads, tails), and we know the average probabilities that result from repeated coin tosses, (50% or 0.5 probability of heads and 0.5 probability of tails). Working with scientists, the policy analyst can define the likely range of outcomes. In many cases, using historical data, models, and results from scientific experiments, the analyst can calculate the probability of the possible outcomes.

If a monetary value of each outcome is known and the probability of occurrence is known, an expected value can be computed to summarize the effect. For example, suppose there is a 5% chance that a toxic spill will occur, resulting in 10 deaths and a 95% chance that a toxic spill will not occur over the life of the chemical plant. Using the valuation methods discussed in Chapter 9, we might calculate that a life is "valued" at $10 million. The expected value is therefore $10 million* 0.05 plus $0 *0.95 or $500,000. This expected value of loss with the chemical plant would be included as one of the costs in a benefit-cost analysis. While risk analysis is much broader than this expected value calculation, it illustrates a best case scenario for policy analysis: the probabilities of each outcome are known, and the monetary value of each outcome is known. As we will discuss later in this chapter, there is still much the analyst can do in a risk analysis if the monetary values are not known.

1.1.2 Uncertainty

In contrast to risk, uncertain situations involve the introduction of a new product or technology where there is little information on what the quantitative probability of each possible outcome is. This situation may be characterized by what we call uncertainty: the alternative states of the world may be known, but we have no basis for calculating the probabilities that each state would occur (Zeckhauser and Viscusi, 1990). If there is true uncertainty, it may be that the policy analysis process must be stopped until scientific experiments are conducted to arrive at these probabilities. In other cases, the analyst might use an expert panel to at least determine which outcomes have a high, medium, or low chance of occurrence. In this case the analysis can display all three of these outcomes for the decision maker and the public to use in its decision process.

1.1.3 Ignorance

Some decisions may be characterized by complete ignorance, where we do not even know enough to list the possible outcomes that might arise, let alone the probabilities (Zeckhauser and Viscusi, 1990). The distinction between uncertainty and ignorance may be illustrated by looking at a couple of real-world issues. Some of the possible outcome associated with continued greenhouse gas emission are known (e.g., rising sea level, warmer temperatures), but the probabilities are not known. Thus climate change might be characterized by uncertainty. The introduction of genetically modified organisms into food crops might be characterized by ignorance: we do not even know what the possible outcomes of such an experiment might be on other plants, animals or humans. Thus this is a situation where there is ignorance. A sound policy, might err on the conservative side – that is, to do no harm – until we at least know what states of the world are likely to result from introduction of these organisms.

It is very important to recognize that risks cannot be eliminated completely. Either as individuals or as a society, we can change the magnitudes of the risks we face – for instance, by eating healthier foods or living on junk food – but we cannot live risk-free. Policy decisions involving risks rarely (if ever) decide that some specific people will (or will not) die. Instead, they change the probability that, on average, some number of people from a population may die earlier or later than they would otherwise. Policy decisions involving risk are, almost by definition, not black and white. Getting that point across to decision-makers and to other affected parties may be your toughest job if you work on risk analysis.

1.2 Decision Trees and Compound Probability of a Series of Events

To make our calculation of expected values a little more realistic, we want to introduce the technique of decision trees. Decision trees allow us to trace out a series of probabilistic events and then compute the expected value of all possible outcomes. In the landfill example above, we might have a 1% or 0.01 probability of a leak and 1 - 0.01 or 0.99 chance of no leak. If a leak occurs, we might have a 0.2 probability that the pollutant ends up contaminating the drinking water of the town. Finally, there may be a 0.4 probability that, if the drinking water is contaminated, people in the town will become ill.

Figure 12-1 below traces out the corresponding branches of the decision tree.

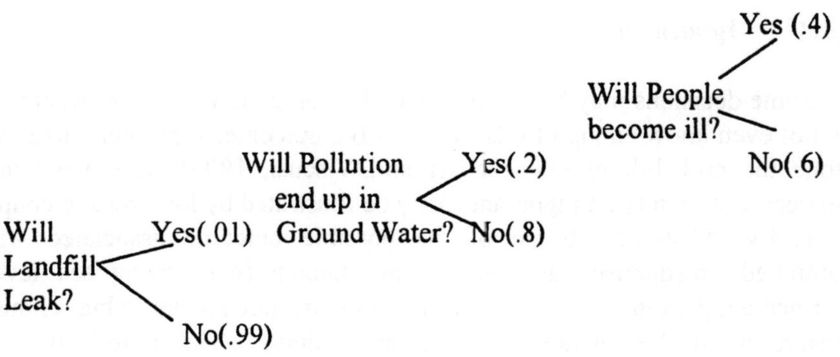

Figure 12-1 Decision Tree and Probabilities

In this simple example, we will say that there are only costs associated with people becoming ill, and ignore any ecological effects. The compound probability that a leak will occur (P = .01), and, if the leak were to occur, that the pollution will end up in the ground water (0.2), and, if the pollution were to end up in the ground water, that people would become ill (0.4) is 0.0008 (0.01*0.2*0.4). As you can see, the odds of all three events occurring together are quite low.

The other piece of information we need to know as part of risk analysis in general, and to calculate the expected value, is how many people will be exposed. For this example, assume the town has 5,000 residents who would be drinking this water. In this example, we would expect that 40 people could become ill (0.0008 * 5,000). Even without the next step of monetary valuation, this information is quite useful to decision makers and the public. This figure says that 40 people would be expected to become ill if the landfill is approved. Of course, the most likely outcome (0.9992) is that no one becomes ill, but the worse case outcome might be that up to 2,000 of the town's 5,000 people would become ill (0.4 * 5,000) if a leak pollutes the city's drinking water. Performing these decision trees helps lay out for the decision maker and the public the range of possible outcomes and their likelihood of occurrence.

If we have calculated that illness would result in lost work and discomfort of, say, $1000, we can go the next step in an expected value calculation. To calculate the expected value, we would multiply the monetary value of damages ($1000) times the 40 people expected to become ill if the landfill is approved. This $40,000 would be the expected cost that would need to be included in the overall environmental policy analysis. For more information on the use of decision tree analysis in environmental policy analysis see Stokey and Zeckhauser (1978).

To see the importance of the risk of exposure information, consider the case if the town considering siting the landfill had 50,000 people. In this case, the expected illnesses would be ten times higher, or 400 people. The corresponding expected costs would also be ten times higher, at $400,000. Developing this type of information can help inform the policy process on the relative merits of different locations for the landfill.

1.3 Some Principles for Modifying Benefit Cost Analysis in the Face of Risk

When projects or policies involve risky outcomes, generally policy analysts compute the expected values of each policy alternative. If the decision-maker and the public are "risk neutral," then selecting the project that has the highest expected value is optimal. By "risk neutral," we mean that individuals are indifferent between receiving $20 for certain and a 50-50 chance of receiving $10 and $30. Thus a risk neutral individual does not enjoy riskier alternatives (i.e., they are not risk loving gamblers who would prefer the 50-50 gamble to the $20). Neither are they "risk averse" individuals who would much prefer the $20 for sure to the gamble, and in fact, might take as little as $18 with certainty rather than the 50-50 gamble of $10 and $30. Thus a risk averse decision maker or society has a larger reduction in utility for a $2 loss than for a $2 gain. If the decision maker or members of society affected by a particular policy decision are risk averse, the analysis must display more than just the expected value of each alternative: it must also provide the distribution of outcomes and their monetary consequences. It may very well be that society prefers to forego a higher expected value for a lower risk.

This should not come as a great surprise to policy analysts. People buy insurance, in part, because they would rather suffer a small loss with certainty (i.e., payment of the insurance premium) than take a risk of facing a large loss. The risk-return trade-off is well known in the investment world as well. Investors must receive higher rates of return that include a risk premium to compensate them for greater chance of an extremely low pay-off or even a loss of capital.

There is a significant amount of literature on modifying procedures and decision rules in benefit cost analysis when the outcomes are risky (Johansson, 1993; Graham, 1981). For example, willingness to pay for a risky outcome is often called option price (Bishop, 1982). Option price is made up of two components: the expected consumer surplus (i.e., the probability of each outcome times the consumer surplus to be realized in that outcome) plus the option value. Option value can be viewed as the additional amount an uncertain and risk averse consumer would pay this time period to ensure the availability of a good in the future – that is, to maintain an option. The risk could either be on the supply side (e.g., will the resource be available next period if I want to consume it?) or the demand side (e.g., will I want to consume the resource in the next period?). The distribution of the risk may also matter. In some market situations, there may be insurance or "contingent claims" markets in which the individuals can hedge their bets, so to speak, against alternative outcomes. However, with many publicly managed natural resources and most environmental settings, no such contingent claims markets exist, and option price is considered the appropriate measure to capture risk aversion of most individuals in a benefit-cost analysis.

2. OVERVIEW OF RISK ANALYSIS

Risk analysis is a much broader approach to incorporating risk into a policy analysis than simply calculating expected values in a benefit-cost analysis. Risk analysis is composed of: (a) risk assessment; (b) risk communication; and (c) risk management.

2.1 Risk assessment

Risk assessment usually involves the quantitative assessment of the damage or injury to society or an ecosystem due to exposure to some agent (e.g., chemical). Risk assessment involves estimating the probabilities of the outcomes, such as increase in cancer deaths in a particular location over a given period of time. If the risks are found to be large enough to warrant concern, then the risk management phase is entered.

2.2 Risk management

Risk management requires the agency to evaluate what control or mitigation options are available – that is, what actions might be taken to reduce this risk. Included in this may be the decision not to approve some action if it is not possible to reduce the risk or if mitigation actions are largely ineffective or have other undesirable effects. A risk management strategy might also include a compensation scheme whereby other risks might be reduced, for instance, by building a hospital in a community without one.

2.3 Risk Communication

If the residual risk levels are viewed as acceptable and the decision is undertaken, then risk communication is conducted to inform the public and aid the public in coping with this risk. This would involve public information programs involving warning labels on the products, emergency notification procedures and available adaptive behaviors that would reduce sensitive individuals' chances of exposure.

2.4 Types of Risk Analysis

2.4.1 Ecological Risk Analysis

There are several types or purposes of conducting the above three steps. For decades, separate risk analyses have been conducted on individual pollutants, chemicals, pesticides, or other risk sources. The emphasis in most of these was on human health effects, such as incidences of illness or life-threatening diseases as such as cancer. Risk analysis is also used in evaluating the performance of engineering and transportation systems, such as aircraft and nuclear power safety. Here the endpoint of interest is human beings.

More recently, risk analysis has been applied to ecosystems in the form of ecological risk analysis. Here the endpoints – the target or focus of the analysis --are risks to non-human species. For example, the toxicity of pesticides to particular species (such as birds and fish) is frequently tested. Ecological risk analysis has seen increasing use in the Endangered Species Act, where the risk of extinction of individual species (e.g., salmon, spotted owls) is being evaluated. Even risks to entire ecosystems might be evaluated in ecological risk assessment. For example, a policy analysis of liberalizing trade in forest products might include the increased risk of introducing exotic insects that could damage forest ecosystems.

2.4.2 Comparative Risk Analysis

Nearly every risk analysis identifies some risk of death and hazards, and there are limited financial resources available to reduce risks. Priorities need to be set as to which are the most important risks to be reduced. We cannot make life risk free. In some cases the opportunity costs of foregoing the use of an otherwise beneficial chemical might be quite large. Technology might be more effective at reducing some risks compared to others. Because of all these factors, priorities need to be set to determine which are the most important risks to be reduced.

Comparative risk analysis attempts to rank the risks relative to one another. It provides a systematic way to identify which are the large risks we should reduce first and which other risks are less demanding of our attention. Usually comparative risk analysis puts a priority on those activities which have the "biggest bang for the buck" – those that reduce the most risk per dollar spent – as a way of minimizing impacts on human health and the environment in the face of limited resources to achieve those goals.

A comparative risk analysis can be performed in a policy analysis by considering not only the risk of the proposed action but also of the alternatives. If the proposed action is to ban a particular use of an insecticide, it is important to consider the risks associated with the substitute insecticide or biological control program – that is, it is critical to consider the "with vs. without" principle. The goal is to avoid piecemeal approaches that may make overall safety worse.

Comparative risk analysis is an attempt to provide some rationality to what can otherwise become a public shouting match over differing priorities. Often individual groups advocate strongly for a particular risk that they face, or for a risk that frightens them more than others. For instance, the risk to an individual due to smoking is, in the U.S., much higher than the risk due to contaminated drinking water; however, because smoking is (more of) an individual choice, while drinking water is harder to avoid, people may advocate reductions in drinking water risks even if anti-smoking activities are much more cost-effective. Comparing safety measures by measuring lives saved per million dollars expended often shows that more mundane risks can be reduced for far less than some headline-grabbing risks. At the same time, objective scientific analysis is not the only basis for public policy. If the psychological costs associated with some risks are much higher to many people than those associated with risks more easily avoided, then policy-makers must weigh the reductions in psychological risk against the reduced number of lives saved.

3. LEGAL BASIS FOR RISK ANALYSIS

In some cases supplementing your policy analysis with features of a risk analysis is often advisable, while in others a risk analysis may be legally required. Many environmental laws contain provisions explicitly requiring a risk analysis as part of the overall policy evaluation. These laws include the Occupational Safety and Health Act (OSHA), Clean Air Act (CAA),

Toxic Substances Control Act (TSCA), Federal Insecticide, Fungicide and Rodenticide Act (FIFRA), the Safe Drinking Water Act (SDWA) and the Food Quality Protection Act of 1996 (FQPA). Some of these acts (e.g., CAA, SDWA) require the agency to choose the alternative that provides the highest degree of safety or ample margin of safety with little consideration of costs. Other acts, including TSCA and FIFRA, require the analysis to balance risks against economic benefits (American Chemical Society and Resources for the Future, 1998). The most recent law, FQPA, requires the U.S. Environmental Protection Agency (EPA) to "...use advanced risk assessment and management methods in dealing with the unique susceptability of infants and children and other vulnerable groups" (Benbrook, 2000:21). Further, EPA is to perform its risk assessments by considering ". . . aggregate exposures to a pesticide from food, drinking water and beverages, in and around the home and public places" (Benbrook, 2000:22). The law provides a more stringent safety test for pesticides used in food consumed disproportionately by infants and children as compared to other foods and agricultural products not consumed (e.g., cotton). In 1994 Congress required the U.S. Department of Agriculture to establish the Office of Risk Assessment and Cost-Benefit Analysis to perform risk assessments of the agency's many programs.

4. STEPS IN A RISK ANALYSIS

4.1 Risk Assessment

Risk assessment involves several steps. First, the source of risk must be identified. Next, individuals (either humans or other species) are exposed to the source. Once these individuals are exposed to the source, the effects of the exposure need to be measured on those individuals. Finally, the effects per individual are applied to the specific population that faces the risks, to determine aggregate effects. Each of these steps involves research based in the natural, physical, biological, and medical sciences; the social sciences and public policy play relatively small roles in these steps.

4.1.1 Source Assessment

A risk assessment begins with an identification of how your alternatives will affect the probability distribution of end-points like health status, injury rates, species extinction. You would sketch out how would modify the existing or future without routine releases of emissions or probabilities of accidental releases, breakdowns, spills, etc. You could use a decision tree approach to trace out the possible outcomes of implementing each alternative in terms of releases, spills, etc. Thus the first step involves tracing the pathways in which the source of risk – for instance, a pollutant or safety measure -- might influence human health or the environment for each alternative. At this stage it is important to work both backward toward the input end and forward to the consumer end in this pathway. This is sometimes referred to as "cradle to grave" or life-cycle accounting. Valuable information about the full costs of an alternative can come from a look at each stage of production, consumption, and disposal. If we are evaluating a new coal-fired power plant versus a nuclear power plant, a complete risk analysis would account for risk to miners, transportation risks of getting the fuel to the plant, emissions and safety at the plant, disposal of waste fuel, construction and decommissioning costs for the plants, and electrical transmissions risks. Ideally, you would estimate or calculate the probabilities associated with each of these effects along the entire pathway.

Figure 12-2 provides a schematic of the five pathways that might be considered in a source assessment for heavy metals. Mining, agriculture, and industries would be the three primary sources of heavy metals, although transport accidents and leaks from landfills and long-term storage may be relevant as well.

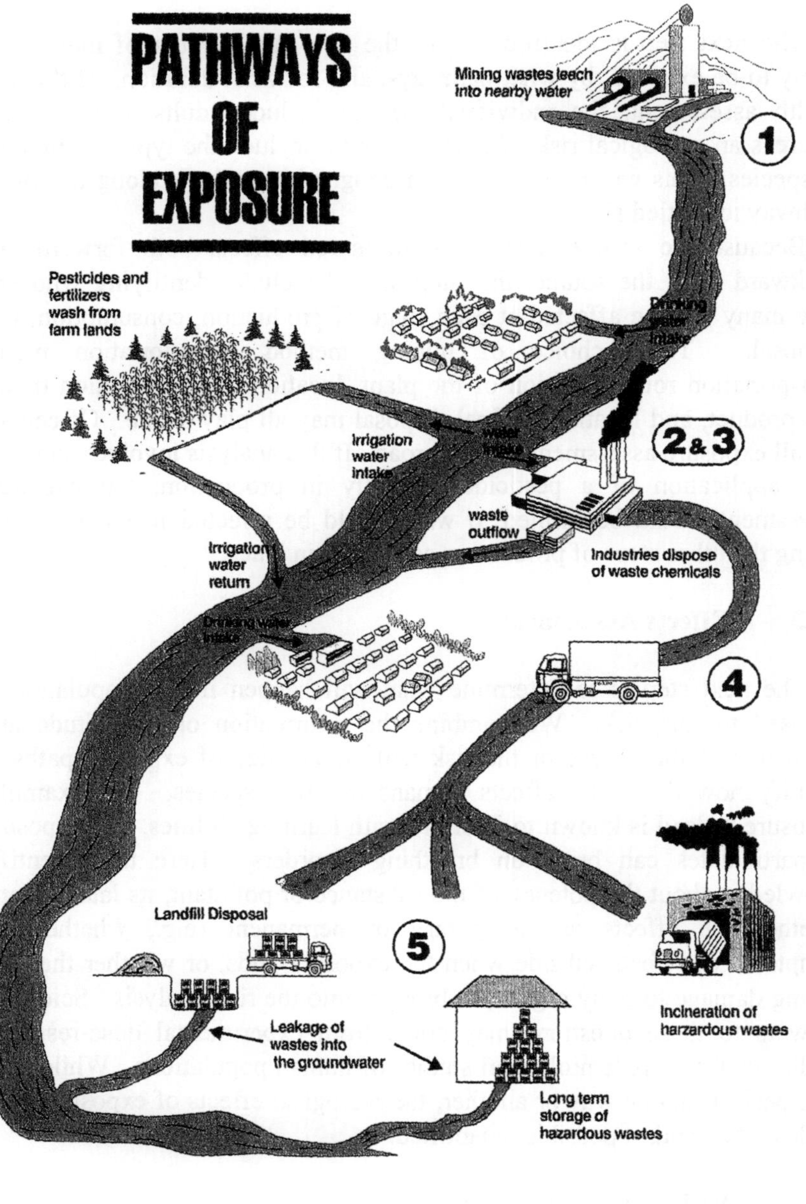

Figure 12-2 Pathways of Exposure

4.1.2 Exposure Assessment

The next step is identification of the type and number of individuals likely to be exposed, by what pathways, and for what duration. If this is a health assessment, the individuals would include adults and children, whereas an ecological risk assessment would include the type and number of species. This will necessarily be a geographic analysis along the entire pathway identified above.

Because the source assessment measured effects both forward and backward from the source, this step should include identifying who and how many will be affected at each stage of production, consumption, and disposal. Thus, choice of mining method, transportation mode, transportation route, location of the plant, location of consumption of the end product, and location of final disposal may all play a role. Of course, not all exposure assessments are so broad. If this analysis involves simply a new application of a pesticide currently in production, the exposure assessment would be limited to who would be affected in the new use, taking the other steps of production as already in place.

4.1.3 Effects Assessment

The next step is to determine what will happen if this population is exposed to this risk. We combine the information on magnitude and frequency of the source of the risk with knowledge of exposure paths to identify how this risk affects humans or other species. For example, exposure to lead is known to interfere with learning abilities, and exposure to particulates can bring on breathing disorders. Here the scientific knowledge about the potency of the substance or pollutant, its latency, and whether the effects are temporary or permanent (e.g., whether the symptoms or effects subside when the exposure ends, or whether there be lasting damage to body organs) is brought into the risk analysis. Scientific answers to these questions may come from experimental dose-response studies or from epidemiological studies of human populations. While each is imperfect, in one way or another, the biological effects of exposure must be described and, if possible, quantified.

4.1.4 Risk Characterization

In this step all of the above information is combined, along with susceptibility of the population, to make predictions about the consequences. That is, the scientific knowledge is combined with the age distribution of the population exposed, its current health status, and availability of medical care, to predict increased incidences of illness, impairment or death. Thus the risk characterization may be different for the same chemical if the group or population segment exposed consists of workers 20-55 years old compared to a population segment that consists primarily of infants, children, and senior citizens. The risk characterization is a description of who is affected, how they are affected, and it often provides a numerical estimate or range of estimates of (e.g.) how many additional cancers or injuries would result if a particular course of action is adopted. The risk characterization should provide documentation as to the procedures used in the first three steps, so that the decision-maker and the public can judge the reasonableness of the resulting numerical estimates. The risk characterization will provide the key information that will be used later in the risk communication step as well.

Figure 12-3 illustrates a simplified flow in the risk assessment:

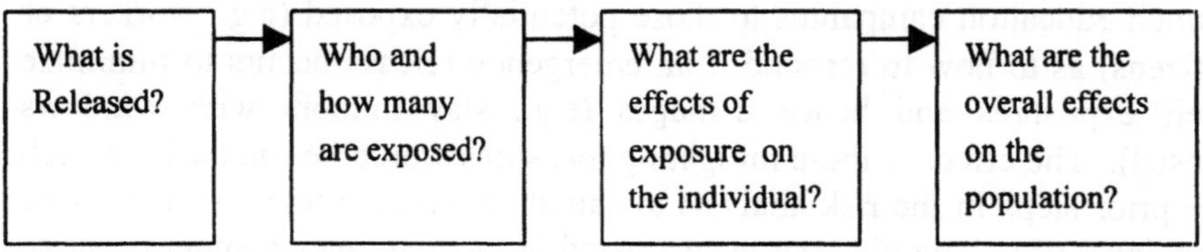

Figure 12-3 Example Risk Assessment Sequence

4.2 Risk Management

Risk management involves determination of whether the estimated risks to the population are acceptable or not, and, if not, what options there may be for reducing the risk. It is usually the judgment of the decision-maker whether the risks are acceptable or worth the benefits that incurring the risk provide. The risk assessment should play a major role, in that it identifies the likely outcomes of the alternatives that the decision-maker faces. The decision-maker often seeks input from stakeholder groups and the public in making this judgment, and she may use cost-effectiveness or benefit-cost analysis in the decision as well. Based on comparative risk assessments, there are sometimes rules of thumb used as to whether a risk is acceptable or not. For example, risks smaller than one in a million (10^{-6}) are sometimes considered small enough not to be of concern (American Chemical Society and Resources for the Future, 1998).

Regardless of the level of risk, if stakeholder groups or the public feel the risk of the action is not worth the benefits, the risk management phase may involve determining whether there are risk-reducing activities that might be included in the proposed action to reduce the risk. This is another place where the risk assessment and the pathways analysis performed above are particularly useful. Re-examining the pathways of exposure may illuminate options for reducing the probability of release, for reducing the number or susceptibility of the individuals exposed, or for mitigating the effects of exposure. This makes the overall risk analysis into an iterative decision making tool. These mitigating measures may include requiring additional backup safety measures, requiring rerouting of trucks carrying waste to highways away from populated areas, or more frequent safety inspections to reduce the likelihood of release, or increased availability of medical examinations of individuals to catch and treat effects early. As a transition to Risk Communication, a risk management strategy may be to launch education campaigns to those potentially exposed (e.g., workers or citizens) as to how to respond if an emergency release occurs to minimize their exposures and hence damages (e.g., stay indoors with windows closed). The effect of these mitigating measures would be run back through the prior steps in the risk analysis to calculate the new level of risk. This new level of risk would then be evaluated in this risk management phase to determine if that risk is acceptable or not. If such a risk is still not acceptable, that may be grounds for not selecting that particular proposed action or alternative (as long as the risk of doing nothing is less than the proposed action) or determining if there is some compensatory project that would reduce other risks so that taking on the new risk is now acceptable. Again, we emphasize here that rarely do people face the choice between risk and no risk; most often, we face (limited) choices over the kinds and magnitudes of risks. Not accepting a risky alternative is likely to leave us facing the current level of risks.

4.3 Risk Communication

This last step involves informing the stakeholders and the public about the risks from the proposed course of action and why these risks were deemed acceptable. The purpose of this step is not necessarily to rationalize the decision, but rather to provide the public with an understanding of what factors were considered in making this trade-off. The discussion should be written in a way that is accessible to the public and provides the context of the decision. Providing comparative risk information about other alternatives considered and everyday hazards is often helpful in this regard.

4.4 Risk Perception Issues

The multiple dimensions of perceived risk often hinder risk communication and public acceptance of objective risk analyses. Whether a given, objectively estimated risk is considered acceptable or unacceptable often hinges on how the risk is perceived by the public. Individuals in society are often more tolerant of risks that they voluntarily choose (e.g., driving, skiing) than those risks that are involuntary (e.g., drinking from a city's water supply, breathing air in a city). In addition to these, the public tends to be more willing to tolerate familiar risks (e.g., driving) than unfamiliar risks associated with a new technology or someone else's level of precautions. Even some deaths are sometimes treated as worse than others, due to the nature of the death or dread (e.g., a slow, painful death from cancer). Finally, there is a well documented certainty premium to eliminate the risk or uncertainty altogether (Zeckhauser and Viscusi, 1990). Thus risk-averse individuals may purposely reject the alternative with the highest expected value for an alternative with a lower expected value, but more certain outcome. While this may be viewed as irrational by those that subscribe to the expected utility view of how individuals should behave, there is much evidence that human behavior does not follow the tenets of expected utility theory (Machina, 1987). Thus, the policy analyst needs to keep these factors in mind when performing a risk analysis.

5. EXAMPLES OF RISK ANALYSIS: HEALTH BENEFIT ASSESSMENTS

Hundreds of risk analyses have been performed over the last 20 years. Examples can be found in Regulatory Impact Analyses performed by U.S. Environmental Protection Agency and other agencies. In this section we wish to present short examples to illustrate the range of application of risk analysis. The details of each of these risk analyses can be found in the citations to these case studies.

The 1983 report on risk assessment by the National Research Council of the National Academy of Sciences serves as a starting point for definitions of risk assessment in EPA (Barry, 1988). Implementation of risk assessment was facilitated by EPA in 1986 when the agency published a series of technical guidance documents for agency staff (Barry, 1988). Much of the initial effort toward risk assessment focused on health effects, especially those due to carcinogens, toxicants, and other chemicals.

The four basic steps EPA followed in many of its risk assessments included:

5.1 Hazard Identification

This involves a qualitative evaluation of whether the available evidence suggests that exposure to this chemical will likely cause any human health effects.

5.2 Dose-Response Assessment

This step attempts to document or estimate a quantitative relationship between concentrations or doses of this chemical in the environment and the frequency of health effects on different types of individuals (e.g., adults, children, asthmatics, etc.) at each concentration. To calculate the likely dose often involves relating emissions and the resulting concentrations in the environment. This step is the same as the "effects assessment" described above.

5.3 Exposure Assessment

This step quantifies the specific types of individuals who are most likely affected (e.g., workers, children), in what concentrations and for what duration from this chemical in this particular policy analysis. All types of exposure are accounted for, including ingestion, inhalation, and contact via skin.

5.4 Risk Characterization

This step relates the dose-response relationship to the exposure assessment to calculate the health effect to the population that will be exposed if this particular policy is adopted. For example, it might estimate the number of additional cancers that can be expected in the population if this particular use of this particular chemical is permitted.

Of course each of these steps is quite involved with literature reviews, models, and (invariably) assumptions to fill in gaps in the literature and data. It is important for the analyst to make these assumptions explicit so that they can form the basis for any sensitivity analysis regarding the uncertain parameters.

This framework has been used by EPA to perform risk assessments for primary health-based standards associated with criteria air pollutants such as carbon monoxide, sulfur dioxide, lead, and nitrogen oxides. Under the Clean Air Act, EPA is to set standards for these pollutants to protect public health with an adequate margin for safety. This does not mean zero risk or zero health effects, but usually the "no observed adverse effect level" —

NOAEL. But providing a margin for safety does mean erring on the cautious side to compensate for incomplete knowledge of how pollution affects human health.

EPA also applies this framework to secondary air quality standards, hazardous air pollutants, and maximum contaminant levels under the Safe Drinking Water Act. In the Toxic Substances Control Act, risk assessment includes both human health effects and the effects on the environment. However, this statute allows for these risks to be balanced with the beneficial uses of the chemical and the availability of substitute chemicals (Barry, 1988). A large risk assessment program for EPA is pesticide registrations under the Federal Insecticide, Fungicide and Rodenticide Act, FIFRA. While this act requires consideration of health and environmental risks, it allows for balancing of economic and environmental costs and benefits. As part of the economic costs, the effects on agricultural production and prices must be considered (Barry, 1988). Pesticide registration has been a contentious area for EPA because of controversy over the appropriateness of weighing risks to people versus cost savings to farmers.

6. CASE STUDY OF ANNULAR DISPOSAL OF OIL DRILLING BRINE WASTES

Oil and gas drilling generally results in large quantities of salt brine by-products. In Ohio, the state of our case study, at least 14 million barrels of brine were generated in 1985 (Hobbs, et al., 1988). Besides containing high levels of salts, the brine may contain toxic contaminants such as benzene, toluene, lead and manganese (Hobbs, et al., 1988). Generally the concentrations of these contaminants in the brine waste are so large that they would have to be diluted with 100 to 4,000 times as much fresh water in order to meet drinking water standards. Since this is not practical, the disposal of brine salts in surface or groundwater essentially makes the water resource unsafe for drinking water purposes.

Due to this contamination problem, most states carefully control the disposal of brine salts from oil and gas production. The safest means of disposal is deep well injection, whereby there are at least three barriers between the injected brine and any freshwater aquifers. However, a cheaper method, called annular disposal, exists. This method injects the brine above the oil producing formation but below any fresh water aquifers. This results in only one barrier between the brine solution and the freshwater aquifer (Hobbs, et al. 1988). Only Ohio and Louisiana allow such practices due to the risk. In the two-year period 1984-1986, 11 water wells were contaminated in Ohio with brine from this annular disposal method. The cost savings to producers of using this approach instead of deep well injection, though, can be large. Since many oil wells produce three or more barrels of brine for each barrel of oil, higher disposal costs can be a burden to many small oil and gas producers in states like Ohio (Hobbs, et al., 1988).

This is a classic environmental policy analysis trade-off: a healthier environment versus higher costs to local producers. Should annular disposal be banned in Ohio as it has in other states? Local citizen groups such as the Ohio Brine Disposal Task Force recommended elimination of annular disposal due to the health risks and availability of the safer disposal options. As you might imagine, industry did not support this, and the legislature had before it a bill to allow the state to continue permitting annular disposal.

Suppose you were the policy analyst in the Ohio legislature assigned to analyze this legislation. How would you approach it? The rest of this case study illustrates how team of researchers at Case Western University approached it. See if you agree or disagree with their approach and their findings. At the end of the case study we will tell you what the legislature decided (no peeking now).

6.1 Risk Analysis Approach Taken

6.1.1 Risk of Contamination

This risk was quantified using the probability of failure during the initial drilling and then probability each year of failure of the barrier and resulting leakage. Using radioactive tracers in similar wells in Louisiana, the respective probabilities were estimated at 0.0025 and 0.009. The next step was to quantify the probability that these leakages would contaminate water wells and then the number of wells contaminated. Conducting this analysis involved using a groundwater solute transport model of the two aquifers most likely affected by annular brine disposal. A given leak was translated in the model into, first, a probability distribution of the contaminant plume, and, second, the plume's movement into the drinking water aquifer. The movement of the plume was traced for twenty years. If, during this time, the plume entered the aquifer and resulted in concentrations of contaminants in excess of drinking water standards, then that area of the aquifer was considered polluted. The size of the plume relative to the size of the aquifer influenced the probability that any particular home water well and city municipal well would be in an area considered contaminated. With the current five-year annular well inspections, the expected number of home water wells contaminated was 2.5. in one aquifer. There was a 0.11 probability that more than 5 home water wells would be contaminated. The probability that a municipal water well would be contaminated was 0.023 in that aquifer as well. In the other aquifer, about one (0.87 probability) home water well was expected to be damaged, and the probability of municipal well contamination was 0.012.

6.1.2 Expected Damages

The methodology used by Hobbs, et al. to calculate damages was not explained as to whether it was based on hedonic property analysis or replacement costs. Nonetheless, they estimated damages of $60,000 per home water well contaminated and $200,000 per municipal water well contaminated. Using these figures and the probabilities, the expected economic damage was $153,000 (roughly 2.5 times $60,000 plus 0.023*$200,000) in the one aquifer and $46,000 in the other aquifer. Thus a total of $207,800 in damages was estimated in the two aquifers.

6.1.3 Cost Savings to Industry

The direct cost savings to industry from being allowed to use the cheaper annular disposal method as compared to deep well injection was estimated on a per barrel basis and then multiplied up by all the current annular wells. The total cost savings were estimated to be $400,000 per year, about twice the damages. This assumes that none of the wells are shut down and all incur the higher deep well injection costs. To the extent a particular oil and gas well is marginal, it may be cheaper to shut down production than to incur the higher costs. If so, the cost savings to industry are slightly overstated. If the analysis used a national accounting stance, there would be no additional benefits beyond the cost savings to industry to consider. (Why is there no effect on oil prices? Think back to the benefit-cost analysis chapter.)

6.1.4 Other Economic Benefits of Annular Disposal

If you were taking a state accounting stance, as these researchers did, you might include employment effects in Ohio. Specifically, you might include the direct and indirect effects on long run employment due to additional drilling of new oil and gas wells in the long run if the cheaper annular disposal option continues to be available. Running these oil/gas employment gains through an input-output model yielded gains in state income of $7.5 million, a sizeable effect. How realistic these gains are is an open question. As discussed in prior chapters, these gains in employment in Ohio are simply transfers of activity that would occur in other oil producing states. However, since this policy analysis was of an Ohio state legislature bill, consideration of these state employment effects were probably relevant. But it may be best not to add them to the $400,000 cost savings from annular disposal.

6.1.5 Benefit-Cost Comparison

On the basis of benefits to be gained from banning annular disposal versus the higher costs to industry, it appears that banning annular disposal would not be economically efficient. Viewed differently, the damages to the oil industry are greater than the damages to homeowners and the city. However, it may be that this analysis did not properly account for inspection costs. As noted below, annualized current costs of inspecting these annular wells every five years is $360,000. If replacing annular disposal with deep well injection reduced the frequency of inspections, this would provide additional benefits from banning annular disposal or reduced cost savings to society from continuing annular disposal.

6.2 Risk Management Options

The current law results in annular brine well inspections only once every five years. More frequent inspections would reduce the likelihood of any brine leaks going undetected. One risk management strategy is to continue to allow annular disposal but to increase the frequency of inspections to once every three years or even once a year.

The authors of the study did address whether the increased frequency of inspection warranted the higher inspection costs. They concluded the higher inspection frequency was not justified. They based this conclusion on the fact that annual inspections would increase the annual inspection costs from $360,000 currently with a five-year inspection cycle to $1.8 million with annual inspection. The reduction in expected number of home wells contaminated did fall from the current 2.5 to 1. But given their estimates of the damages to households from contaminated wells ($60,000), they conclude that increased frequency of inspection is not worthwhile. Given their damage estimates to households, one would have to agree: there is a $90,000 damage reduction or benefit to reducing the number of expected contaminated wells from 2.5 to 1, whereas the increase in costs is nearly $1.5 million.

The risk management option suggested by the authors was to compensate households who had their water supplies contaminated. This could be done by purchasing them replacement water, for instance, and using a separate water storage system such as a cistern coupled with water trucked in at the oil company's expense.

As part of the risk characterization, the authors noted their assumption was that there was continuous monitoring of the drinking water wells, and that any detection of contamination would result in immediate shutdown of the drinking water wells (Hobbs, et al., 1988). This of course may not always be realistic, and at best the expense of such a system needs to be deducted from the cost savings to industry from annular disposal. If the authors estimated damages to households at $60,000 under these optimistic assumptions, then health damages could be much larger with more realistic monitoring assumptions. Further, the authors were quite explicit that their analysis did not account for any effects on fish, wildlife, plant communities, or other ecological effects. Leaks that travel through one of the aquifers eventually discharge into a river. If some of the contamination is discharged into the river, it would lead to direct damages to riparian vegetation, fish, and waterfowl and indirect damages to fish-eating birds and anglers. Given the authors' estimated cost savings of $200,000 a year, the fisheries, waterfowl and angling benefits would not have to be large to offset the cost savings. Further, if the optimistic assumption of continuous monitoring of drinking water wells and immediate notification is replaced with more conservative assumptions about drinking water well notification, the damages may be larger. Nonetheless, this case study provides a nice simple example of a risk analysis and some potential pitfalls in conducting even a simple risk analysis.

6.3 Legislative decision

The legislature went on to allow permitting of annular disposal, but placed a few technical restrictions on it. What do you think of this decision, given the analysis above? How might you have refined or improved upon this analysis?

7. ECOLOGICAL RISK ANALYSIS OF USDA CONSERVATION PROGRAMS

In 1994 the U.S. Department of Agriculture was required by Congress to establish an Office of Risk Assessment and Cost-Benefit Analysis to perform internal reviews of the agency's programs. The Department of Agriculture's first assessments of national programs were of the Conservation Reserve Program (CRP) and the Environmental Quality Incentives Program (EQIP). CRP is designed to retire highly erosive land from agricultural production to reduce soil erosion, reduce non-point source water and air pollution (e.g., dust), as well as provide some limited wildlife habitat benefits. EQIP is water quality-oriented assistance to farmers to reduce non-point source pollution and salinity of rivers via activities such as modifying agricultural practices or planting buffer strips. To conduct risk assessments of these environmental programs, the Department of Agricultural adapted EPA's (EPA, 1992, 1996) ecological risk assessment framework (Powell and Wilson, 1997).

There were three steps in the risk assessment: (a) problem formulation; (b) analysis; and (c) risk characterization. The problem formulation phase involved selecting endpoints that would be evaluated in the analysis. The endpoints for CRP included water quality, wetland function, and wildlife habitat (Department of Agriculture, 1997a). The endpoints for EQIP included potable water supplies, wetlands, viability of aquatic communities, and threatened or endangered species (Department of Agriculture, 1997b). The analysis phase involved characterization of the exposure of endpoints to agricultural effects and measurements of each of the endpoints with the respective programs in place. For example, the analysis involved linking how the program would reduce the amount of nitrates running off farm fields into water bodies, how the nitrates affect water quality, and how water quality affects aquatic communities such as trout (Powell and Wilson, 1997). The risk characterization phase combines the exposure and effects analyses to provide the magnitude, severity, spatial, and temporal distribution of the adverse effects on each of the endpoints. For USDA, the results of the risk characterization would be used with the economic analysis and other administrative factors in arriving at an overall evaluation of the performance of these two programs and how they might be improved.

Resources for the Future's review of the Department of Agriculture's efforts (Powell and Wilson) details the difficulties in performing ecological risk assessment. Some endpoints were difficult to measure in a quantitative fashion. For example, wetlands and viability of aquatic communities were vaguely defined, making measurement of the endpoints difficult (Powell and Wilson). Operationalizing these concerns using commonly employed measures of species diversity would have at least made possible (although no doubt very costly to implement) measurement of the endpoint for aquatic communities (Powell and Wilson). Some of the other endpoints that were measurable lacked definition of thresholds of concern that would affect the risk characterization. For example, little use of lethal dose measurements such as LD50, the dose whereby half the population would die, was used as a threshold indicator. Without such quantitative indicators it is difficult to tell the level of effect at which a resource become impaired or non-functional. While there are thresholds for soil erosion in terms of tons per acre, there were few well-defined thresholds for other endpoints (Powell and Wilson).

The point of reviewing this risk assessment is to provide some illustration of the steps involved in an ecological risk assessment and the difficulty of performing ecological risk assessment relative to human health risk assessments. However, do not think that, because it is hard to do, ecological risk assessments should be avoided. On the contrary, it is only by attempting these that progress is made. Data gaps such as those identified in the Department of Agriculture's ecological risk assessment help direct research to fill those data gaps. There is also a clear "learning by doing" improvement that occurs every time an assessment like this is performed. The next one is likely to be better.

Chapter 13

Examples of Policy Analyses

While we have provided examples of actual policy analyses throughout the earlier chapters of this book, this chapter provides three in-depth analyses. We do not have the space to show you applications of all the tools and methods discussed in this book; instead, we have chosen these to demonstrate aspects that are important and commonly used.

The first example illustrates the U.S. Forest Service's use of a decision matrix that incorporates the criteria discussed in Chapter 3. These matrices are commonly found in Environmental Impact Statements prepared by nearly all federal and state agencies. This example highlights the transition from general criteria to measurable performance indicators, and it also shows how useful it can be to assemble comparable information on all alternatives into one format.

The second example illustrates the application of benefit-cost analysis for comparing market values of hydropower with non-market values of recreational viewing of a large waterfall. There is a direct trade-off between the amount of water visible pouring over the falls and the amount going through the pipeline to the hydroelectric turbines: more water over the falls means less water for electricity. While electricity has an obvious economic value, recreational enjoyment of the sights and sound of the waterfall have a less easily observable value. In this example, the economic benefits of viewing the falls at different river flow rates is compared to the economic value of the hydropower to determine the economically efficient level of water for the falls and hydroelectricity production. This example illustrates both the use of benefit-cost analysis and the incorporation of non-market effects in the analysis.

The third example illustrates how optimization can be applied to evaluate the impact on profits and nitrate leachate of alternative best management practices to reduce non-point source water pollution from farmland. The optimization approach allowed for comparison of several economic and non-economic regulatory approaches and varying levels of stringency of each. This example demonstrates how important modeling the effects of alternatives can be: some quite reasonable-sounding alternatives ended up both reducing profits and increasing pollution, outcomes that this study could help avoid.

As a policy analyst you will often spend as much time reviewing and critiquing other analysts' studies as actually performing your own policy analyses. Thus, as you read each of these case studies, think about how else these analyses might have been done. In particular, what were the primary decision criteria that were analyzed, and what criteria were omitted? Think about the simplifying assumptions that were made. Ask yourself what additional type of data could have been collected to reduce the reliance on assumptions. Think about whether other analytical decision techniques could have been applied to analyze these problems. These are the types of questions you should be thinking about at this point in the book.

1. DECISIONS USING MATRICES: AN EXAMPLE FROM U.S. FOREST SERVICE EIS

Most Environmental Impact Statements (EISs) summarize the physical, biological, economic and social effects using a matrix. As discussed in Chapter 3, the general form of the matrix has the criteria and associated performance indicators as a column along the left side, with the alternative policy actions along the top row. The matrix provides a concise way of comparing the effects of the alternatives on a variety of criteria of importance to a decision-maker.

Table 13-1 presents an example of an especially thorough decision matrix from the U.S. Forest Service (USFS) for the Siuslaw National Forest, which covers the Coast Range of Oregon. The Siuslaw National Forest is a coastal forest with a high stream density which contain important anadromous fish species, such as salmon and steelhead. The Forest also has one of the highest values of timber, on a per acre basis, of any National Forest. The U.S. Forest Service is required by law to develop a management plan for the Siuslaw National Forest for the next 50 years; Chapter 11 includes a description of the Multiple Use-Sustained Yield Act requirements for how the USFS should manage these forests. Table 13-1 illustrates the types of performance indicators for biological, social, administrative, and economic criteria in forest planning.

The matrix also illustrates the cross-walk from planning issues to the decision matrix. The USFS needed to take general areas of concern and translate them into measurable criteria for the matrix. The key public and agency resource issues, and the indicators chosen to represent them, were:

1. How much timber will be harvested? This general concern was translated into three performance indicators: the long term sustained yield of timber, number of acres harvested, and the allowable sale quantity (ASQ) of timber each year.

2. How much of the existing old-growth forest stands will be maintained? The performance indicator is the acres of existing old growth stands maintained by the end of the plan (i.e., 5th decade).

3. What quantity and quality of anadromous fish habitat will be provided on the Forest? The public, the fishing industry, and the state fish and game agency are interested in maintaining productive fish habitat on the five (out of Oregon's seven) coastal anadromous fish streams contained within the Forest. The actual performance indicator used is Salmon Habitat Index.

4. How much habitat will be provided for wildlife and threatened and endangered species, and how and where will these habitats be managed? The public's concern focused on elk and adequate habitat for viable populations of old growth-dependent species, such as the spotted owl. The performance indicators include numbers of elk and number of pairs of spotted owls and their associated habitat areas.

5. What diversity of recreation opportunities will be provided? The public is primarily concerned about Forest Service supply of semi-primitive non-motorized (SPNM) recreational opportunities such as hiking and backpacking. The performance indicator is the percentage of future SPNM recreation demand met.

6. How much of the Forest will be managed as Special Interest Areas (SIA)? Special Interest Areas can be designated by the Regional Forester to protect unusual scenic, historic, scientific or other special use values. The public has expressed an interest in having more SIAs on the Forest. There are three potential areas for SIA designation on the Siuslaw National Forest. Various alternatives will contain additional acreage for SIAs, with acreage and number of SIAs used as performance indicators.

7. Which areas will be managed to maintain or enhance visual quality or scenic resources? Landscapes seen from highways, rivers, and developed recreation areas are called viewsheds. Protecting scenic quality of viewsheds (by eliminating timber harvests or modifying timber harvest layout) can contribute to an enjoyable visit to and through the Forest. The performance indicator is a visual quality index.

8. How will management of Forest resources affect economies of local communities? National Forest outputs provide a base for lumber and other wood products, commercial fishing, and tourism. In addition, a portion of receipts from the sale of Forest outputs (primarily timber) is paid to county governments surrounding the Forest. Two performance indicators are jobs and payments to local governments.

9. What economic value will the Forest resources generate in the future? There is both a national and local concern about the economic value of the Forest outputs. There is also a concern over the size of the Forest budget and returns to the U.S. Treasury from the Siuslaw National Forest. The three performance indicators are net present value (NPV), Forest Service budget, and receipts.

1.1 Development of Alternatives

The National Environmental Policy Act (NEPA) requires that an agency consider a wide range of alternative means to address issues. First, NEPA requires that an EIS have a continuation of current management direction (the no action alternative); this reflects the without-project condition, or baseline. The regional office of the U.S. Forest Service provided two themes for development of additional alternatives: one emphasizing market/commodity outputs (e.g., timber), and another emphasizing nonmarket/amenity outputs (e.g., wildlife and dispersed recreation). Remaining alternatives were formulated to address the nine specific issues listed above. Some of the differences between alternatives were manifested in terms of how much land would be available for allocation to each management emphasis (old growth, scenic viewsheds, timber harvesting, etc.), and some differences reflected the management practices that would be applied (i.e., different means to attain the ends or address the management issues).

The following is a list of the eight alternatives and the theme or intent of each.

A. Alternative A is continuation of the existing management. It reflects future levels of wood production, elk, and bald eagles expected with continuation of existing management practices.

B. Alternative B stresses wood production with other natural resources only produced at levels that would meet the minimum legal requirements.

C. Alternative C emphasizes economically efficient production of wood and big game habitat (particularly elk) and a variety of recreational opportunities. Elk population would increase by a factor of three by the fifth decade of the planning period.

D. Alternative D is the market emphasis alternative. Emphasis is given to wood, commercial fish such as salmon, and developed recreation activities for which a fee is paid.

E. Alternative E emphasizes economically efficient wood production and provision of a variety of recreational opportunities and wildlife habitats. The key feature is longer timber rotations. This alternative eventually became the preferred alternative.

F. Alternative F would emphasize recreation and protection of scenic resources as well as habitat for fish and nongame wildlife. Only "moderate" amounts of timber would be produced.

G. Alternative G gives emphasis to production of nonmarketed multiple uses including water quality, fish, wildlife, dispersed recreation, and scenic resources. Nearly all of the land suitable for timber would be managed on long rotations (60% with rotations of 100 years or more).

H. The goal of alternative H is to preserve natural systems in large areas of the Forest, protect nongame wildlife habitats, and provide maximum protection to municipal watersheds.

Table 13-1 presents the matrix that was developed by the U.S. Forest Service to compare how each alternative did on addressing the public issues.

Table 13-1. Summary of selected indicators by alternative

Criterion	Current	A*	B	C	D	E**	F	G	H
TIMBER 1ST DECADE	n/r[1]	73	75	72	66	68	58	30	21
Allowable Sale Quantity	n/r	69	71	68	62	62	55	30	20
Acres Harvested	n/r	6750	6340	6329	5821	5909	4990	2766	1836
OLD GROWTH STANDS									
1000 Acres Maintained, 5th Decade	n/r	17	16	18	16	19	22	34	34
FISH									
Salmon Habitat Index	598	851	845	895	952	911	1000	1172	1202
WILDLIFE									
Elk (5th Decade)[2]	3000	4880	6000	12560	2430	7470	2050	5500	760
Spotted Owl Pairs (1st & 5th Decade)	39	37/26	37/26	37/26	37/26	37/32	39/39	39/39	39/39
Spotted Owl Habitat Areas	22	40	40	40	40	46	53	62	72
SPECIAL INTEREST AREAS									
No. Areas	1	2	2	3	2	4	3	4	3
1000 Acres	2	3	2	5	1	6	7	7	5
RECREATION									
%SPNM Demand Met in 50 Years	n/r	24	31	54	27	54	70	83	71
VISUAL RESOURCE									
Visual Quality Index (100 is high)	n/r	30	7	22	9	35	40	62	71
EMPLOYMENT									
1000 Jobs, 1st Decade	8	9	10	10	9	9	8	6	5

Criterion	Current	A*	B	C	D	E**	F	G	H
PAYMENT TO LOCAL GOVERNMENTS									
$Millions, 1st Decade	12	17	18	18	16	16	14	8	5
RECEIPTS									
$\$10^6$	46	70	73	70	64	64	57	31	20
BUDGET									
$\$10^6$ 1st Decade	26	20	28	28	26	26	22	17	17
NPV@4%[3]									
($Billions)	n/r	2	2	2	2	2	2	1	1

*Continuation of existing management direction alternative.
**Preferred alternative.
[1]Not reported.
[2]Oregon Department of Fish and Wildlife (ODFW) goals are 10,000 elk.
[3]Monetary value of benefits and costs discounted to the present time at a 4% discount rate.

1.2 Summary of Table 13-1

The list of alternatives provides a wide spectrum of future "visions" of what the Siuslaw National Forest could be. Each of these alternatives was assessed according to the key issues to determine what contributions they made toward addressing those issues or resolving particular concerns.

The quantitative predictions of each of the performance indicators for each alternative demonstrates a truly a remarkable effort by U.S. Forest Service personnel. The interdisciplinary team of foresters, hydrologists, fisheries and wildlife biologists, recreation planners, and economists utilized a wide variety of models to make these predictions. They also worked with state resource management agencies in this analysis.

What is apparent from this table is that consistent information is available on biological factors (timber, fish, wildlife), social factors (recreation, visual quality, employment), administrative feasibility (budget), and economics (present net value). This ensures that a decision-maker can evaluate how each alternative does on all of the decision criteria and in addressing the policy issues and that the decision will be based on a clear understanding of the relative effects of the alternatives.

1.3 Selection of Preferred Alternative

The Forest Supervisor and Regional Forester selected Alternative E as the preferred alternative in the Draft EIS. This alternative does provide a balance or compromise among the different resources and criteria. That is, Alternative E is neither heavy commodity-oriented nor heavily preservation-oriented. Further, the preferred alternative attempts to avoid reductions of the level of key resources below their current level. For example it provides elk, salmon, semi-primitive nonmotorized recreation, jobs, and payments to local counties at or above current levels without increasing the budget. While other alternatives may increase one or two resources more than the preferred alternative, the trade-off is reductions in other resources below their current levels. Reductions below the current level sometimes bring forth a firestorm of protests by those that view themselves having *de facto* property rights to the current level of resources they use. Thus it is sometimes a good strategy to minimize the reductions below the current level, even if it means giving up some of the gains above the current level. It also appears from the selection of Alternative E as the preferred alternative that the U.S. Forest Service treated each evaluation criterion equally. In addition, Alternative E appears fairly responsive to the nine public issues the Forest Plan sought to address. This matrix illustrates how a carefully constructed table can summarize a great deal of biological, social, and economic information that highlights the differences across the alternatives.

2. COMPARING RECREATION VALUES TO HYDROPOWER IN HYDROPOWER RELICENSING

2.1 Statement of the Problem

There are thousands of privately owned hydroelectric facilities in the United States. Although these facilities provide electricity using the renewable hydrologic cycle rather than non-renewable fossil fuels, and produce no air pollution, these facilities are not without environmental impacts. Hydropower facilities often change the flow regime of the river, frequently to the detriment of native fisheries. In addition, hydropower facilities that divert water into penstocks reduce the flow of the river between the point of diversion and point of return, often adversely affecting recreational uses and the aesthetics of the riverine environment.

Concern over the environmental impacts of hydropower was a factor addressed by Congress when it passed the Electric Consumers' Protection Act of 1986. In this Act, Congress required the Federal Energy Regulatory Commission (FERC) to give "equal consideration" to non power or environmental values in its licensing or relicensing decision. The relicensing decision is usually related to how much water should be left in a stream for fish and recreation. Often times the more water left in the stream, the less that goes through the turbines. This results in a significant opportunity cost to the utility, and ultimately to electricity customers.

There has usually been a strong difference of opinion between natural resource agencies and the dams owners over what river flow represents equal consideration and balancing. Without commensurate units to compare kilowatts and visitor days, FERC can only make subjective judgments. The purpose of this example is to illustrate the application of benefit-cost analysis and non market valuation to a FERC relicensing decision involving determination of economically optimum flows over a scenic falls on a major western river. While economic approaches should not be the sole criterion for making relicensing decisions, at present little comparable information on benefits and costs is provided to FERC decision makers or to the public. The result has been a preoccupation with setting minimum instream flows rather than optimum instream flows that balance costs and benefits.

2.2 Determining the Economic Benefits of Greater Flows

Benefit-cost analysis (BCA) puts both the benefits of increased flow and the opportunity costs of hydropower into commensurate units, dollars. Federal BCA guidelines (U.S. Water Resources Council 1983) require that economic value be measured as net willingness to pay (WTP). The net WTP for hydropower is often approximated by the replacement cost of the next least expensive source or the costs avoided by reliance on hydropower (Gibbons, 1986). For viewing the falls, the net WTP is measured by a visitor's consumer surplus. The change in recreation benefits resulting from a change in flow is composed of two parts (Duffield, et al. 1992:2171). The first part is the change in visitation associated with the change in flow. When flows are low the waterfalls are less attractive and fewer people will visit and vice versa, when the flow is higher. The gains or losses in trips (T) are then valued using a WTP function. The second part is the change in value per day of those people continuing to visit the falls at different flow levels. As shown by Duffield, et al. 1992, the change in total recreation values (RV) equals:

(13-1) $\partial RV/\partial Flow = [(\partial T/\partial Flow) * WTP(flow)] + [\partial WTP/\partial Flow * T(Flow)]$

2.3 Benefit Methodology

To provide an estimate of how WTP changes with flow (the second term in equation (1) we used the contingent valuation method (CVM). CVM involves surveying the visitors about their willingness to pay a higher amount (e.g., trip costs) to view the falls at higher water levels. To estimate the first term in equation (1), a variant of CVM, called contingent behavior, is used to obtain how visitors' trip frequency would change with flow conditions.

2.3.1 Visitor Survey Sample

The relevant survey population is the 57,000 visitors to Shoshone Falls which is located in a designated recreational area in Idaho. The falls are a well-known attraction in the region and easily accessible from an interstate highway. The falls are over 200 feet high and 800 feet wide at maximum flows. At high flows the falls provides an impressive sight and sound. The names and addresses of visitors to the falls were collected via an on-site survey. At total of 1500 surveys were sent out. Each mailed survey included an individually addressed cover letter. The survey package included a postage paid return envelope. A follow-up postcard was sent two weeks following the initial mailing.

2.3.2 Survey Design

The survey format, visual aids, and initial questions were designed jointly with a social science consulting firm to elicit not only WTP and changes in trip taking but also visitor preferences for different flows. Portions of the survey were pre tested with focus groups in Idaho and Washington. The focus group subjects were drawn from the recreation site visitor mailing list. The survey format, question order, and question wording were extensively revised following these sessions. Next, a complete draft of the revised survey was mailed to another group of individuals, also drawn from the on-site sample. Using a telephone interview, these individuals were then guided through the survey to uncover problems in interpretation of the questions that might lead to lack of face validity of the survey.

The final survey was short (4 pages long) and consisted of two main sections: (1) questions regarding visitors most recent trip; (2) visitor preference ratings, WTP and trips associated with four color photos that depicted the falls at 50 cfs (photo C), 250 cfs (photo B), 790 cfs (Photo A) and 2,000 cfs (Photo D). The range of flows was meant to include the current minimum, possible flows that could be adopted by the FERC and a large enough range that a statistical WTP and trip function could be estimated to allow analysis of any flows within this range. Effort was made to neutrally present the views of the falls at the four different flows and not to influence the respondent's own assessment of the desirability of one particular flow over another. To facilitate comparison of what the falls looked like at each flow level, all four photos were presented on one page.

The wording of the question asking how visitors would change the number of trips taken in response to different flows over the falls was: *The appearance of the Falls might influence your decision to make the trip to the Falls. If you could be certain that the Falls looked like photos A, B, C or D, how would it affect the additional number of trips to the Falls you would make each year?* The categories for each photo were More Trips, Same Number of Trips, Fewer Trips and Wouldn't Go. For the categories, More Trips and Less Trips the respondent was requested to write in the number of additional trips or the number of fewer trips.

The wording of the WTP question was "*The appearance of the Falls might also influence how much you would spend to get the opportunity to view the Falls. If you could spend money to be certain that the Falls would look like photos A, B, C or D, how would it affect the* <u>*additional*</u> *amount your party would have spent to make the trip to the Falls on your last visit?* The categories for each photo were patterned after a simplified payment card (Mitchell and Carson 1989) having +\$15, +\$10, +\$5, 0, -\$5, -\$10, -\$15, as well as allowing for a fill-in-the-blank of more dollars or less dollars. The range from positive to negative was needed because some of the photos reflected flows lower than what the visitors had experienced on their most recent trip and hence they might only be willing to visit the site if it costs less. Since the WTP question asks for "...the additional amount your <u>party</u> would have spent...", to calculate per person WTP, we also asked party size.

2.4 Survey Results

The survey response rate was 63% after deleting undeliverable surveys. This response rate is quite reasonable given that no second mailing of the survey was performed (to keep costs to a minimum) and that most respondents were on multi-destination or multi-purpose trips. This response rate resulted in 886 useable surveys.

2.4.1.1 Statistical Analysis

The general form of the WTP regression is given in equation (13-2):

(13-2) WTP per person per day= f(Flow, Demographics)

The general form of the visitation function is similar:

(13-3) Change in trips per person = f(Flow, Demographics)

Increased flow over the falls is expected to have a positive benefit and economic theory suggests diminishing marginal value of additional flows. That is, each additional 100 cfs of flow would most likely have a smaller and smaller incremental value, other things remaining equal. A non-linear functional form allowing for diminishing marginal value is the natural log of flow as measured by CFS. This natural log functional form appears consistent with a plot of WTP as a function of flow on the most recent trip.

The results of the regressions indicate that the natural log of flow is statistically significant in both the WTP per day equation and the change in number of trips equation at well beyond the 0.01 level. The flow variable is important because it is the key policy variable in this analysis.

2.5 Aggregate Annual and Monthly Benefit Estimates

The annual benefits relation with flow was calculated following the formula in equation (13-1): The change in number of trips regression, and specifically, the regression coefficient on flow provides our estimate of ($\partial T/\partial Flow$). This change in number of trips per visitor was applied to the base visitation for each geographic area to estimate the level of aggregate visitation at each flow level. This level of visitation was valued at the WTP per visit associated with that specific flow. Reductions in number of trips below current use (i.e., lost trips due to low flows) were valued at the average value per trip during the base or survey season. These two sources of the change in benefits were then summed as required by equation (13-1). To calculate the marginal benefits curve in Figure 13-1, we calculated the change in aggregate benefits by expanding from the sample to the known total visitation, divided by the change in CFS.

2.6 Opportunity Costs of Hydropower Foregone

The benefits of increased flows must be compared to the change in the opportunity cost of hydropower foregone before any conclusion can be made about the economic optimum flows. In essence, the marginal benefit of water over the falls must be compared to the marginal opportunity costs. Although the exact opportunity costs will vary from utility to utility depending on their replacement power opportunities, estimation of approximate hydropower values is relatively straightforward because standard factors are available. Each cfs of water passing a fixed point for 1 day amounts to 1.9835 acre feet of water. Therefore, 100 cfs of flow per day represents 198 acre feet of water. An acre foot of water generates .87 kilowatt hours (kwh) per foot of head (Gibbons 1986:89). Because the study site's hydroelectric plant has 207 feet of head, this translates into 180 kwh per acre foot of water. If we require 100 cfs bypass the powerplant and instead go over the falls during the 12 daylight hours per day for the 6 month recreation season, this requires 18,068 acre feet of water and would sacrifice 3,252,240 kwh of electricity.

The value of hydropower is the cost avoided by the utility over the next least expensive source. This varies from a national average of $.017 per kwh using coal fired steam electric plant (Gibbons 1986: 92) to $.0192 per kwh in the Pacific Northwest (Hamilton and Whittlesey 1988). For the utility under study, the weighted average of six days per week at peak power (during the daytime) and Sunday off-peak is $.01653 per kwh. Applying this kwh value, the opportunity cost of foregone hydropower is $53,759 for 100 cfs or $538 per cfs. Because the energy production and value of electricity per unit are constant across the range of flow alternatives being considered, the opportunity cost per acre foot is constant and hence the marginal cost curve in Figure 1 is horizontal.

2.7 Determining Economically Optimum Flows

Figure 13-1 plots the marginal benefit of each additional cfs of flow against the opportunity costs of flows. If just one optimum flow must be selected for the entire recreation season, then Figure 13-1 indicates the economic optimum flows to maintain just the aesthetic value of viewing the falls (assuming no other benefits such as fisheries and riparian vegetation) is about 235-240 cfs during the recreation season when both value per day and number of trips vary with flow. This is the optimum because the marginal benefit of additional flows just equal the marginal cost at 235 cfs. For flows greater than this level the marginal benefit curve is less than the marginal cost curve, indicating total net benefits (benefits minus costs) are falling. Review of U.S. Geological Survey gauging station flows for the segment of the river above the turbine intakes and the falls indicates that there are adequate flows in every month, even during drought years to meet this 235 cfs requirement.

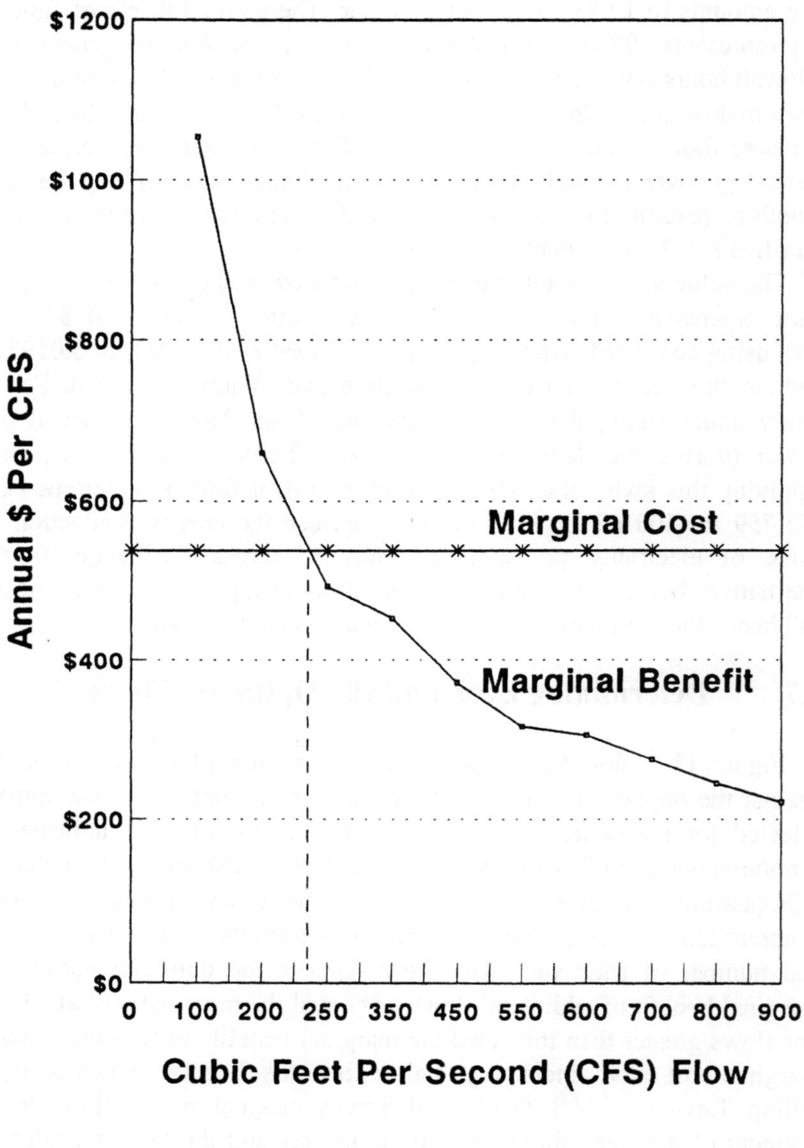

Figure 13-1. Comparison of Marginal Benefit and Marginal Costs of River Flows

2.8 Conclusion

This example has demonstrated that non power values such as the aesthetics of water flowing over a falls can be measured in dollar terms using carefully developed and implemented surveys along with appropriate statistical analysis. This type of analysis can provide a quantitative approach to including environmental values that are commensurate with hydropower values. The approach illustrated in this example suggests it would be relatively straightforward for the FERC to balance benefits and costs when attempting to give "equal consideration" to non power values as required by the Electric Consumers' Protection Act of 1986. Although non market valuation cannot be reduced to some "cookbook" table of values, a site specific economic analysis would make performing trade-offs more objective as is illustrated in Figure 1, where benefits and costs of alternative flows are graphically compared. At present, fisheries biologists and recreation planners are often at a disadvantage when all they have are narrative statements or measures of weighted usable area of fish habitat and utilities have dollar opportunity costs of power foregone. Although economics should not be the decisive factor in public policy decisions, being able to quantify most of the affected benefits and costs on a comparable basis would contribute to ensuring that non power resources really are given equal consideration in FERC licensing and relicensing decisions.

3. REGULATING NON-POINT SOURCE WATER POLLUTION

3.1 Statement of the Problem

Agricultural runoff – including soil erosion, fertilizer, and pesticides -- is a major source of water pollution. It typically is a "non-point" source of pollution, which means that there is no one easily identifiable point where the pollution runs into the receiving body. Instead, agricultural runoff typically moves across the surface of the land or through sub-surface soil and arrives at a lake, river, or aquifer without an easily observable entrance point. Regulating non-point source pollution (NPSP) is much more difficult than regulating point sources of pollution for two reasons: first, because of its diffuse nature, it is very difficult, if not impossible, to measure the amount of the pollution; secondly, because it can move over long distances, it is difficult or impossible to identify the source of the pollution. Nevertheless, because of the importance of this problem, regulators at both the state and national levels are expressing growing interest in reducing NPSP.

Farms can reduce their runoff through use of "best management practices" (BMPs). BMPs are farming practices, such as changing tillage or applications of fertilizers or pesticides, that reduce pollution from farmland. What is a "best" management practice depends on the crop, on the particular location (for instance, the soil type), on the nature of the environmental problem, and on the economics of the practice. Identifying the most effective ways to reduce pollution in a farming area requires an understanding of the relationship between farm practices and farm profits as well as the relationship between farm practices and pollution for that location.

Taylor, Adams, and Miller (1992) sought to identify the effects of different policies to reduce NPSP in the Willamette Valley of Oregon. The Willamette Valley has highly diversified agriculture, with products including grass seed, grains, vegetables, and berries. The valley also experiences high levels of nitrogen and phosphorus in its surface water and groundwater, two nutrients that contribute to eutrophication of water bodies (growth of algae that sucks all the oxygen out of the water and kills fish) and pose health risks to people as well. They sought to identify ways to reduce agricultural runoff that would not create unnecessary hardship on farmers in the area.

3.2 Methods: Setting Up the Programming Problem

First, because BMPs are crop- and geography-specific, they separated their study area into five different farm types: well drained bottomland, poorly drained bottomland, well drained terrace land, poorly drained terrace land, and well drained foothills. Next, they identified the farming activities that can be used on each of these land types. These farm practices ranged from existing practices to pollution-reducing BMPs.

The third step required estimating the effects of the different farming activities when applied to the different land types. As is common in environmental policy analysis, Taylor, Adams, and Miller needed to draw on the expertise of the sciences in conducting their analyses. To estimate these effects, they used an agronomic model, the Erosion-Productivity Impact Calculator (EPIC). EPIC simulates how a plant grows based on environmental inputs (including soil type and climate) and management practices (including timing and type of fertilizer applications or different cropping systems). Its output includes both crop yield and nutrient runoff.

By running EPIC to simulate the effects of crop growth over 25 years, they were able to determine average productivity instead of the outputs of just one year. EPIC was used both to simulate the effects of current farm practices as well as the effects of various BMPs. While EPIC does not perfectly reflect the effects of existing practices, the authors expected that the differences in the practices would be more accurate. Since they were concerned mostly with the changes due to use of BMPs, instead of the absolute levels, the biases in EPIC were not expected to be serious flaws in their analysis.

Finally, their project was intended to examine the effects of different proposals to reduce runoff. They considered five different regulatory approaches: a tax on nitrogen runoff; a tax on fertilizer; per-acre effluent standards; requiring the use of no-till drills on small grains and grass seed production; and banning fall fertilizer applications, to reduce runoff during winter rains.

The results of the EPIC analyses were crop production and runoff associated with an acre of a specified kind of land put in a specified management practice. As discussed in Chapter 11, these kinds of results can feed very easily into a linear programming model. The activities in this case are the choices of land management practices (which include both the choice of crop and the practices used to grow the crop) applied to each acre of each type of land. The technical constraints provide the amount of the crop grown and the amount of runoff from each acre in each activity. The policy constraints were, sequentially, the different regulatory approaches that they considered. Finally, their objective function came from the assumption that farmers maximized profits. Implementing the objective function was possible with additional information on crop prices and the costs of each management practice.

Their linear programming problem (in simplified notation) was to

(13-4) Maximize Price*Yield – (Cost per unit of input)*(Input levels) – (Tax per unit of runoff)*Runoff

(note that the tax was only included in the objective function for the scenario where runoff was taxed). The objective function included the sum of revenues and costs for all crops on all farms.

The constraints linked the activities – the management practices associated with each type of acre – to yield, input levels, and runoff. Thus, for instance, Z_k represented each acre put into practice k. Total yield of crop i was yield per acre from practice k growing crop i times Z_k, the number of acres of that practice. Inputs were associated with each acre of Z_k as well, as was runoff per acre of Z_k. Constraints also limited the total acres of each land type. Finally, the effects of each management practice were simulated by imposing them as additional constraints on the system.

3.3 Analysis of the Baseline and of Policy Alternatives

First, they simulated crop production and runoff in the absence of any regulatory approach to NPSP. As we argued back in Chapter 2, the effects of any policy can only be measured against a baseline. Additionally, as Taylor, Adams, and Miller argue themselves, although there might be biases in baseline results due to use of EPIC, these biases are likely not to affect changes due to the regulatory approaches. That baseline is presented as the first results row in Table 13-2

They then simulated a variety of policies that a government agency might consider. For instance, restricting average leachate below a specified level is equivalent to restricting total leachate; those scenarios produced the combination of management practices that would maximize profits while reducing pollution below a specified level. Taxing leachate is, from an environmental economics perspective, an efficient way of achieving the same goal as restricting average leachate. Because leachate is not directly measurable, though, neither of those efficient policies would be likely to be implemented in practice. Taxing fertilizer, on the other hand, would discourage farmers from applying the nutrients that ended up as NPSP and would be relatively easy to implement. Restricting leachate per acre, another difficult practice to implement, simulates the effects of requiring heterogeneous farmers to all achieve the same level of runoff per acre (and is predicted to be inefficient). Finally, requiring no-till drills and a fall fertilizer ban are specific, observable management practices that might influence pollution.

The programming model, as described in Chapter 11, not only gave them the optimal set of management practices in response to each of the regulatory programs, but it also served the accounting practice of providing information on resulting effluent and profits per acre and in total. Table 13-2 provides some of the results that they present, which indicate some very interesting differences across some of the policies. For instance, taxing leachate produces the same allocation of land management practices as "least-cost" policies (those presented in the table as average leachate less than some level), but it has much higher costs for farmers, who, in addition to changing management practices, must pay the tax on remaining leachate. (Compare, for instance, reducing average leachate below 7.5 lbs./acre, which results in a runoff reduction approximately equivalent to a $4/lb. charge on leachate, but which costs about $35/acre less). This result is completely expected, but it highlights a disadvantage of the use of pollution taxes.

Perhaps surprisingly, a 50% tax on nitrate fertilizer leads to less than a 2% change in nitrate runoff while imposing a 17% reduction in profits. Helfand and House (1995) found a similar result: farmers seem unwilling or unable to reduce runoff when faced with fertilizer taxes. Also perhaps surprising, two practices intended to reduce runoff – using no-till drills and banning fall fertilizer applications – actually led to increased leachate. The former appears to be due to the nature of the practice: while it reduces erosion, it does increase nitrate leaching. In the latter case, farmers were predicted to substitute to different crops that did not require fall fertilizer applications. These different crops in fact increased runoff. Thus, an important advantage of conducting these policy simulations is not only that better practices can be found, but also those practices with undesirable effects can be avoided.

Table 13-2. Profits and runoff changes due to NPSP policies (summarized from Taylor, et al., 1992)

	Profit Per Acre		Nitrate Leachate Per Acre	
Policy	$	Percent Change	Lbs.	Percent Change
Unrestricted	147.33	--	16.6	--
Ave. Leachate < 12.5 lbs.	135.99	-7.7	12.5	-24.9
Ave. Leachate < 7.5 lbs.	117.21	-20.4	7.5	-54.9
Ave. Leachate < 1 lb.	88.65	-39.8	1.0	-94.0
Tax $4/lb. of Leachate	87.23	-40.8	7.6	-54.5
Tax $12/lb. of Leachate	77.93	-47.1	0.7	-95.9
50% Tax on Fertilizer	122.05	-17.2	16.4	-1.7
Leachate < 20 lb./acre	133.56	-9.4	12.8	-23.0
Leachate < 10 lb./acre	90.60	-38.5	1.6	-90.5
Leachate < 6 lb./acre	88.48	-39.9	0.8	-94.2
Requiring No-Till Drills	144.87	-1.7	20.8	+25.0
Fall Fertilizer Ban	137.03	-7.0	18.5	+10.8

3.4 Conclusion

The results show clearly that different policies have different effects, and that adverse effects can happen, either on farm profits or on environmental impacts, if a policy is not chosen carefully. Especially because restricting or taxing leachate directly may not be technically feasible, indirect policies, such as fertilizer taxes or requiring specific practices, may at first appear to be desirable and easily implemented alternatives. This kind of simulation effort can be extremely valuable both in showing the effectiveness of these alternatives and in helping regulators to develop more effective approaches to reducing non-point source pollution.

4. SUMMARY

These examples all suggest that a typical policy analysis is often limited in the scope of the decision criteria that are addressed and the analysis techniques applied. Perhaps due to these features, they were able to complete the analysis and illustrate the trade-offs between alternatives. Also notice that, as more decision criteria are incorporated, such as in the U.S. Forest Service matrix, the "best alternative" often becomes far less obvious. With narrower decision criteria, such as the benefit-cost analysis of hydropower versus recreation, the analysis appears to be more clear-cut. We say *appears to be*, as we don't want you to assume that the precise recommendation on economically efficient flows in hydropower example is the only decision rule that a policy-maker will actually use. While fewer the decision criteria make it easier to identify the "best" alternative, can you say that that alternative really is the best? Perhaps not, when the other decision criteria are brought into play. Nonetheless, we hope these examples illustrate that you can make a contribution to the policy process without analyzing every last detail, from every possible viewpoint, while employing commonly used analytical techniques that we have provided you.

Chapter 14

Concluding Guidance

Back in Chapter 4, we reminded you that you had to present your results and conclusions to your client. In many ways, for the purposes of this book, you have been our client. Here is our last chance to leave you with the results and conclusions that we hope you will take away from this book.

We have presented you both with general principles of policy analysis and with a number of specific tools that are often useful in conducting analysis. If you need to use a particular tool, we encourage you to keep this book around and look up the details, or go to some of our references and learn even more. The following discussion will, we hope, reinforce the large-scale issues that you are likely to face as an analyst.

1. GENERAL PRINCIPLES

Your job as an analyst is to provide your client with information. Your client can then choose how she wishes to use that information. Even if your analysis suggests to you the best possible course of action, it is typically not your job to make that decision. Policy analysis is an important tool for decision-makers: it can be influential in their choice of alternatives, and it can be a useful means to sway others' opinions as well. Your analysis is unlikely, though, to be the decision rule, the sole determinant of the best alternative. Your client may want to include factors that were not incorporated in your study, or her priorities may have changed since she commissioned your study and the original priorities may no longer be her current priorities (it happens!). If you are an economist, by training your priority may be economic efficiency, while your client may be a public official who places more weight on distributional impacts. You should keep in mind this distinction between a policy analysis as decision tool and analysis as a decision rule: you provide the tool, but your client will most likely make the rule!

One of the most important conclusions that you can take from this book is that policy analyses can and do take many forms. The effects of environmental policies range from ecological to economic, from sociological to scientific, and include just about any other discipline you can name. As a policy analyst, you could develop results in one set of impacts (for instance, the ecological effects), or you could synthesize, as we discussed in Chapter 3, all the impacts into one grand matrix. No one form of policy analysis is the right one for all situations.

There may well be a right form of policy analysis for a specific situation and a specific client, though. Your job as an analyst is to do the work that will most benefit your client. She may need information in order to form an opinion on a particular issue, or she may want a study that she can use to support her preconceived position. As Chapter 1 discussed, there are clear ethical concerns that can arise in your position. The most important principle that you can have and that you can relay to your client is that you will let the analysis results fall where they may. You should *never* change your results because they do not support your employer. Instead, you and your client can influence the outcome of your study by your choice of analysis topic, and you can discuss together what will become of your results if they do not provide the "right" answer. Your client can benefit from your results even if they provide bad news; she may change her mind, or she may develop new solutions to the problem. Honesty is the best policy if you want to maintain and enhance your reputation as an analyst.

2. GUIDANCE FOR DOING THE ANALYSIS

Policy analyses can take many forms, from sketches on the back of an envelope to multi-year, multi-million-dollar studies, and covering a wide range of disciplines. Still, some common principles underlie any analysis in which you might be asked to participate.

First, always keep in mind the "with-without principle" (Chapter 2) and identify the baseline for your analysis. What would the world be like in the absence of the policy you are studying? If you cannot identify the baseline, how can you possibly determine the effects of the policy? It is highly tempting to compare the effects of a policy to the world *before* the policy, because we know how the world before the policy looked. The world, however, is a constantly changing place. You have to be able to distinguish between changes that would have taken place anyway, and changes that are due to the policy. If desirable (or undesirable) changes would take place regardless of enactment of the policy, then the policy will have no effect, and you need to tell that to your client. If your preliminary analysis indicates that a particular resource or level of effect will be the same regardless of the policy alternatives, state this in your report, and then you do not need to focus any further on it. Choices between alternatives are driven by effects that vary across alternatives, not factors that are the same regardless of the alternative.

Secondly, you should use an analytical tool (or set of tools) that is appropriate for the analysis you are conducting. If your client is interested in the local economic impacts of a policy, conducting a benefit-cost analysis from a national stance is inappropriate. A regression model uses different data and frames the analysis differently than an optimization model (which indicates a single best resource allocation, given the constraints). Any of these can be useful in a wide number of settings, but none of them is appropriate in all settings. You should have enough understanding of the range of tools that exists that you can find the appropriate one for your study. If the right tool isn't in your tool kit, your job is either to learn the right tool or to find someone else who knows how to use it. Also keep in mind that you may well need to use multiple tools for your analysis. A benefit-cost analysis, for instance, may require use of an ecological processes model, an epidemiological model, some statistical regression, and perhaps even an optimization model. For all these reasons, you should be aware of the wide range of models and methods that might be of use to you.

Third, you should approach all data with some skepticism. If you have ever tried to measure anything (your weight, a cup of flour, setting the time on a watch), you know that measurement error occurs, to varying degrees, in almost all situations. In addition, data that you might use in your analysis may have been collected with a completely different purpose in mind. The sampling scheme for the data may not be the right one for your study, or the data might measure something close to but not exactly what you need, or the data might be grouped differently than you would like (e.g., by county rather than city). No data set is perfect. You should try to get and use the best data available to you, but recognize that getting better data is not always worth the time and effort involved.

Fourth, as we discussed in Chapter 4, you should use an iterative approach in conducting your analysis. As you analyze your data, new alternatives may occur to you. You may find that you really need to identify better data or use a different model than you originally planned. There may be decision criteria that you had not previously included but need to be brought to the attention of your client. While policy analysis is very applied research, it is research nonetheless; and, like any research activity, it is likely to raise as many new questions as it answers. While you need to keep to the scope (Chapter 2) on which you and your client agreed, you may also want to propose additional projects after you complete the one at hand. The key is to keep your client informed of your progress, and you may need to ask for additional guidance if you have to substantially revise your original study plan. It is often critical to produce some preliminary results within the required time frame rather than constantly ask for more time. At some point a policy decision will be made, with or without your results; the policy process may not wait for you. If you feel you have any valuable information to contribute, it is important to get this to your client in a timely manner. While your preliminary results may not be perfect, it may represent a quantum leap in a decision-maker's understanding of the intended and unintended consequences of a policy. Remember that the best is often the enemy of the better.

Fifth, it is important in the design and implementation of your policy analysis to emphasize the distinction between the means and the ends. Sometimes your client will not only know what she wants to accomplish but also the means by which to do it. As an analyst, you need to recognize that there are usually several alternative means to attain a given end or objective. Often the disagreement in the policy process is not over the ends or goals (which are often shared by many of the stakeholders) but by the means to attain them. For example, most everyone agrees that protecting human lives is good. But business and environmentalist will disagree on the mechanism to achieve this goal. Should we reduce air or water pollution, or should we promote smoking avoidance programs or personal fitness training? Often what is most effective in attaining the goal is an empirical question. This empirical evidence is just what a good policy analysis should provide. But remember to separate out the end goals from the means. Failure to do that has been the downfall of many policy analyses and real policies!

Related to this point is confusion in the policy process between what is an input and what is an output or result. For example, acres allocated to wildlife habitat is an input to producing a sustainable population of certain species. Generally, what people care about is the output or result, and less about the details of how it is done. Often, resource specialists think the output or end result is acres of wildlife habitat. But protecting additional acres of wildlife habitat may be a necessary but not sufficient condition to increase wildlife populations. Failure to account for predation by non-native species or pesticide drift into the wildlife habitat may result in the wildlife habitat not actually supporting a self-reproducing population of the desired species, even if sufficient acres are protected. People get enjoyment and benefits from seeing or photographing (and sometimes harvesting) the animals, not the habitat per se. So always keep this distinction in mind in dealing with resource managers. What they view as a desirable end state may just be an intermediate step to producing the end result that is of value to people.

Sixth, be realistic about the time and monetary budget you will need to perform your policy analysis. An inexpensive "quick and dirty" analysis is often all that is appropriate to address some minor policy issues. While everyone recognizes that you could do a more thorough and in-depth study, the values at risk do not always justify spending a great deal on the study. Evaluating a decision about whether to install a new stoplight at an intersection should not cost $100,000, as it would be cheaper just to put the stoplight in. Conversely, a decision about whether to build a $50 million sports stadium to attract a professional sports team would probably warrant a $500,000 study, not a $10,000 quick-and-dirty analysis. You and your client will almost need to do a simple benefit-cost analysis on whether more detail and precision will improve the decision process. They key is to tailor the level of your analysis to the magnitude of the consequences of the policy.

Lastly, read the literature and stay current in this field. This book cannot do justice to all the creative ways in which policy analysis can and will be conducted in the future. There are entire journals, including the *Journal of Policy Analysis and Management, Risk Analysis,* and *Policy Sciences,* that publish policy case studies and present new methods for policy analysis or how existing methods can be adapted for performing a policy analysis. Government web sites, such as that of the U.S. Environmental Protection Agency, often contain hundreds of completed regulatory impact analyses that can illustrate different approaches. We encourage you to attend professional meetings, such as those of the Society for Risk Analysis. Seeing how others perform a policy analysis can give you insights useful for future policy analyses. If your client will let you present your policy analysis to others, do so. Obtaining constructive criticism from your colleagues is a great way to improve your skills. Most of all, both the authors learned what little we know about policy analysis by doing it. If we can do it, so can you!

3. THE ROLE OF POLICY ANALYSIS IN THE POLICY PROCESS

So, after all that hard work and those late nights to complete your analysis for your client, what will actually happen to all your work? We would like to tell you that your client will thank you profusely and make her decision immediately based entirely on your analysis. As honest policy analysts ourselves, though, we should warn you not to expect this outcome.

Policy analysis, probably more often than not, does influence how your client thinks about your issue and what alternative is chosen. That influence, though, may be indirect and unobservable and may only take place over a long period of time. Your client may take a long time to read your results, or she may not understand them very well (again, talk to her about how the results should be presented). The timing on your report may be wrong – she may have had to make the decision before you finished, or delays in the process may mean that the decision will not happen for a long time. Other considerations may have arisen in the meantime that make your results less important. You may have produced results that do not support her position, and she may choose to ignore your analysis. Or, it may take her a long time to digest the results and be influenced by them.

Do not be discouraged by any of these outcomes. Influencing minds is a gradual and evolutionary process. You may need to explain to your client not only what your results are, but also how she can use them. You may need to explain multiple times why the alternative that she prefers is dominated by another one. If your results are available to others, you may be able to influence others. Most decision-makers are intelligent people, even if they are often overworked and lacking in time to understand all the complexities of any one issue. If you gain her trust through credible, useful analysis, you are likely to see your results show up in policy decisions.

REFERENCES

Adamowicz, W., J. Louviere and M. Williams. 1994. Combining Revealed and Stated Preference Methods for Valuing Environmental Amenities. Journal of Environmental Economics and Management 26:271-292.

Adamowicz, W., J. Louviere and J. Swait. 1998. Introduction to Attribute-based Stated Choice Methods. Final Report Submitted to the Damage Assessment Center, National Oceanic and Atmospheric Administration, U.S. Department of Commerce. Washington DC.

Adams, Darius and Richard Haynes. 1980.The 1980 Softwood Timber Assessment Market Model. Forest Science 26(3) (supplement).

Aillery, M., P. Bertels, J. Cooper, M. Moore, S. Vogel and M. Weinberg. 1996. Salmon Recovery in the Pacific Northwest: Agricultural and Other Economic Effects. Report #727. USDA Economic Research Service, Washington DC.

Alberini, A. and A. Krupnick. 1997. Air Pollution and Acute Respiratory Illness: Evidence from Taiwan and Los Angeles. American Journal of Agricultural Economics 79(5): 1620-1624.

Alward, Gregory and Charles Palmer. 1983. IMPLAN: An Input-Output Analysis System for Forest Service Planning, in IMPLAN Training Notebook, Land Management Planning, Rocky Mountain Forest and Range Experiment, U.S. Forest Service, Fort Collins, CO.

Alward, G., H. Davis, K.Despotakis and E. Lofting. 1985. Regional Non-Survey Input-Output Analysis with IMPLAN. Paper presented at Southern Regional Science Association Conference. Washington DC.

Alward, Gregory, William Workman and Wilbur Maki. 1992. Regional Economic Analysis for Alaskan Wildlife Resources, in Valuing Wildlife Resources in Alaska, edited by G. Peterson, C. Swanson, D. McCollum and M. Thomas. Westview Press, Boulder, CO.

American Chamber of Commerce of Mexico. 1997. NAFTA's Success: A Three Year View from Mexico; Executive Summary. July 1. http://web.lexis-nexis.com.

American Chemical Society and Resources for the Future. 1998. Understanding Risk Analysis. Washington DC.

American Lung Association. 1996. Study Finds New Evidence of Ozone's Effect on Lungs. Press Release, November 21.

American Lung Association. 2000. American Lung Association Fact Sheet: Children and Ozone Air Pollution September. http://www.lungusa.org/air/children_factsheet99.html.

American Lung Association. 1997. Air Pollution Cleanup Could Boost Economy, Says New Lung Association Report. Press Release, May 13.

Arctic Power. 2000. Top Ten Reasons to Support Development in ANWR. http://www.anwr.org/topten.html.

Arrow, K., R. Solow, P. Portney, E. Leamer, R. Radner and H. Schuman. 1993. Report of the NOAA Panel on Contingent Valuation. Federal Register 58(10):4602-4614.

Arrow, Kenneth J. 1965. Criteria for Social Investment. Water Resources Research 1: 1-8. Reprinted in Economics of the Environment: Selected Readings, Third Edition, Robert Dorfman and Nancy S. Dorfman, editors (New York, W. W. Norton and Company, 1993).

Barron, Eric J. 1995. Global Change Researchers Assess Projections of Climate Change. Eos 76(18) (May 2): 185, 189-90.

Barry, T. 1988. An Overview of Health Risk Analysis in the Environmental Protection Agency. In Risk Analysis and Management of Natural and Man-Made Hazards, edited by Y. Haimes and E. Stakhiv. American Society of Civil Engineers. New York, NY.

Benbrook, Charles. 2000. Apples, Kids and Core Science. Choices, 3rd Quarter: 21-24.

Bishop, Richard. 1982. Option Value: An Exposition and Extension. Land Economics 58(1): 1-15.

Bjornstad, David, Ronald Cummings and Laura Osborne. 1997. A Learning Design for Reducing Hypothetical Bias in the Contingent Valuation Method. Environmental and Resource Economics 10: 207-221.

Boadway, Robert. 1979. Public Sector Economics. Winthrop Publishers, Cambridge, MA.

Boyle, K., J. Poor and L. Taylor. 1999. Estimating the Demand for Protecting Freshwater Lakes from Eutrophication. American Journal of Agricultural Economics 81(5):1118-1122.

Boyle, Kevin and John Bergstrom. 1992. Benefit Transfer Studies: Myths, Pragmatism and Idealism. Water Resources Research 28(3):657-663.

Bradley, Iver and James Gander. 1968. Utah Interindustry Study: An Application of Input-Output Analaysis. Utah Economic and Business Review 28(2).

Brookshire, David, William Schulze, Mark Thayer and Ralph d'Arge. 1982. Valuing Public Goods: A Comparison of Survey and Hedonic Approaches. American Economic Review 72(1):165-177.

Brown, Thomas and Gregory, 1999. Why the WTA-WTP Disparity Matters. Ecological Economics 28(3):323-336.

Carrothers, T., J. Graham and J. Evans. 1999. Valuing the Health Effects of Air Pollution. Risk in Perspective 7(5): 1-6.

Carson, R., N. Flores, K. Martin and J. Wright. 1996. Contingent Valuation and Revealed Preference Methodologies: Comparing Estimates for Quasi-Public Goods. Land Economics 72(1):80-99.

Carson, R. et al. 1997. Temporal Reliability of Estimates from Contingent Valuation. Land Economics 73(2):151-163.

Carson, Richard, Theodore Groves, and Mark Machina. 1999. Incentive and Informational Properties of Preference Questions. Department of Economics, University of California, San Diego.

Champ, P., R. Bishop, T. Brown and D. McCollum. 1997. Using Donation Mechanisms to Value Nonuse Benefits from Public Goods. Journal of Environmental Economics and Management 33(2):151-162.

Coppedge, Robert. 1977. Income Multipliers in Economic Impact Analysis, #400 X-5. Cooperative Extension Service, New Mexico State University, Las Cruces, New Mexico.

Cropper, M. and M. Freeman. 1991. Valuing Environmental Health Effects. In Measuring the Demand for Environmental Quality, J. Braden and C.Kolstad, eds. Elsevier, Amsterdam.

Cummings, R., D. Brookshire and W. Schulze. 1986. Valuing Environmental Goods: A State of the Art Assessment of the Contingent Valuation Method. Rowman and Allenheld, Totowa, NJ.

d'Arge, Ralph and Jason Shogren. 1989. Non-Market Asset Prices: A Comparison of Three Valuation Approaches. In H. Folmer and E.van Ierland, eds. Valuation Methods and Policy Making in Environmental Economics. Elsevier, Amsterdam.

Davis, Lawrence and Norman Johnson. 1987. Forest Management, Third Edition. McGraw-Hill, New York, NY.

Davis O. and A Whinston. 1965. Welfare Economics and the Theory of Second Best. Review of Economic Studies 32: 1-14.

Desvousges, W. H., F. R. Johnson and H. S. Banzhaf. 1998. Environmental Policy Analysis With Limited Information: Principles and Applications of the Transfer Method. Northampton, MA: Edward Elgar.

Desvousges, William, Michael Naughton and George Parsons. 1992. Benefit Transfer: Conceptual Problems in Estimating Water Quality Benefits Using Existing Studies. Water Resources Research 28(3):675-683.

Dickie, Mark. 2000. Averting Behavior and Cost of Illness Methods. Draft mansucript, Department of Economics, University of Southern Mississippi.

Duffield, John, John Loomis, and Rob Brooks. 1987. The Net Economic Value of Fishing in Montana. Montana Department of Fish, Wildlife and Parks, Bozeman, MT.

Duffield, J., C. Neher, T. Brown. 1992. Recreation Benefits of Instream Flow: Applicatlon to Montana's Big Hole and Bitterroot Rivers. Water Resources Research 28(9):2169-2181.

Dwyer, John, John Kelly and Michael Bowes. 1977. Improved Procedures for Valuation of the Contribution of Recreation to National Economic Development. Water Resources Center Report 128. University of Illinois, Urbana-Champaign, IL.

Feldstein, Martin. 1985. International Trade, Budget Deficits and the Interest Rate. Journal of Economic Education 16: 189-193.

Foster, H. and Bruce Beattie. 1979. Urban Residential Demand for Water in the United States. Land Economics 55(1):43-58.

Freeman, M. 1993. The Measurement of Environmental and Resource Values: Theory and Methods. Resources for the Future, Washington, D.C.

Fredricks, Kenneth, Tim VandenBerg and Jean Hanson. 1996. Economic Values of Freshwater in the United States. Discussion Paper 97-03, Resources for the Future, Washington, D.C.

Gan, C. and E. J. Luzar. 1993. A Conjoint Analysis of Waterfowl Hunting in Louisiana. Journal of Agricultural and Applied Economics 25(2): 36-45.

Garrod, G. D. and K. G. Willis. 1997. The Non-Use Benefits of Enhancing Forest Biodiversity: A Contingent Ranking Study. Ecological Economics 21:45-61.

Gerking, Shelby, Menno DeHann, William Schulze. 1989. The Marginal Value of Job Safety: A Contingent Valuation Study. Journal of Risk and Uncertainty 1:185-199.

Gibbons, Diana. 1986. The Economic Value of Water. Resources for the Future, Washington, D.C.

Goodstein, Eban. 1995. Benefit-Cost Analysis at the EPA. Journal of Socio-Economics 24(2) , Summer:375-389.

Graham, D. 1981. Cost Benefit Analysis Under Uncertainty. American Economic Review 71: 715-725.

Gray, Lee and Robert Young. 1984. Valuation of Water on Wildlands. In Valuation of Wildland Resource Benefits, edited by G.Peterson and A. Randall. Westview Press, Boulder, CO.

Hamilton, J. and N. Whittlesey. 1988. Markets for Stored Water in the Pacific Northwest. Unpublished Report. Moscow Idaho: Department of Agricultural Economics, University of Idaho.

Hanemann, Michael. 1984. Welfare Evaluations in Contingent Valuation Experiments with Discrete Responses. American Journal of Agricultural Economics 67(3):332-341.

Hanemann, Michael. 1989. Welfare Evaluations in Contingent Valuation Experiments with Discrete Response Data: Reply. American Journal of Agricultural Economics 71(4):1057-1061.

Hanemann, Michael. 1991. Willingness to Pay and Willingness to Accept: How Much Can They Differ? American Economic Review 81(3):635-647.

Hausman, Jerry. 1981. Exact Consumer Surplus and Deadweight Loss. American Economic Review 71(4):662-676.

Hecker, Jayetta Z. 1997. Statement of Jayetta Z. Hecker, Associate Director National Security International Affairs Division, International Relations and Trade Issues, U.S. General Accounting Office. Testimony before the House Committee on Ways and Means, September 11. http://www.house.gov/ways_means/trade/105cong/9-11-97/9-11heck.html

Helfand, Gloria and Brett House, 1995. Regulating Nonpoint Source Pollution Under Heterogeneous Conditions. American Journal of Agricultural Economics 77: 1024-1032.

Helfand, Gloria and James Peyton. 1999. A Conceptual Model of Environmental Justice. Social Science Quarterly 80(1):68-83.

Herriges, Joseph and Cathy Kling, eds. 1999. Valuing Recreation and the Environment. Edward Elgar Publishers, Northampton, MA.

Hewitt, Julie and Michael Hanemann. 1995. A Discrete/Continuous Choice Approach to Residential Water Demand Under Block Rate Pricing. Land Economics 71(2):173-192.

Hobbs, B., C. von Patterson, J. Heslin and M. Maciejowski. 1988. Annular Disposal of Oil and Gas Brines: The Risks and the Benefits. In Risk Analysis and Management of Natural and Man-Made Hazards, edited by Y. Haimes and E. Stakhiv. American Society of Civil Engineers. New York, NY.

Hufbauer, Gary Clyde, Daniel C. Esty, Dianna Orejas, Luis Rubio, and Jeffrey J. Schott. 2000. Announcement of NAFTA and the Environment: Seven Years Later. October. http://www.iie.com/publications/toc.cfm

Johansson, Per-Olov. 1993. Cost-Benefit Analysis of Environmental Change. Cambridge University Press, Cambridge, England.

Jones-Lee, M., M. Hammerton, R. R. Phillips. 1985. The Value of Safety: Results from a National Survey. Economic Journal 95:49-72.

Julius, Brian 1999. Discounting and the Treatment of Uncertainty in Natural Resource Damage Assessment. Technical Paper 99-1. Damage Assessment Center, National Oceanic and Atmospheric Administration, Silver Spring, Maryland.

Just, Richard, Darrell Hueth and Andrew Schmitz. 1982. Applied Welfare Economics and Public Policy. Prentice Hall, Englewood Cliffs, NJ.

Kennedy, Peter. 1998. A Guide to Econometrics, 4th edition. MIT Press, Cambridge MA.

Kopp, Raymond. 1992. Why Existence Value Should be Used in Cost-Benefit Analysis. Journal of Policy Analysis and Management 11(1): 123-130.

Knestch, Jack and Jack Sinden. 1984. Willingness to Pay and Compensation Demanded. Quarterly Journal of Economics 99: 507-521.

Lareau, T. J. and D. A. Rae. 1989. Valuing WTP for diesel odor reductions: an application of contingent ranking technique. Southern Economics Journal 55(3): 728-742.

Leggett, Christopher G., and Nancy E. Bockstael. 2000. Evidence of the Effects of Water Quality on Residential Land Prices. Journal of Environmental Economics and Management 39(2):121-44.

Little, Jane Braxton. 1995. The Quincy Library Group. American Forests 101 (January/February): 22-24.

Lind, Robert. 1990. Reassessing the Government's Discount Rate Policy in Light of New Theory and Data in a World Economy with a High Degree of Capital Mobility. Journal of Environmental Economics and Management 18(2): S8-S28.

Lipsey, R. and K. Lancaster. 1957. The General Theory of Second Best. Review of Economic Studies 24(1):11-33.

Loomis, John. 1988. Broadening the Concept and Measurement of Existence Value. Northeastern Journal of Agricultural and Resource Economics 17(1):23-29.

Loomis, J. 1989. Test-Retest Reliability of the Contingent Valuation Method: A Comparison of General Population and Visitor Responses. American Journal of Agricultural Economics 71(1):76-84.

Loomis, John. 1992. The Evolution of a More Rigorous Approach to Benefit Transfer: Benefit Function Transfer. Water Resources Research, Volume 28(3), March 1992 PAGES.

Loomis, J. 1993. An Investigation into the Reliability of Intended VisitationBehavior. Environmental and Resource Economics 3:183-91.

Loomis, John. 1993. Integrated Public Lands Management: Principles and Applications to National Forests, Parks, Wildlife Refuges and BLM Lands. Columbia University Press, New York, NY.

Loomis, John. 1996a. Measuring the Economic Benefits of Removing Dams and Restoring the Elwha River: Results of a Contingent Valuation Survey. Water Resources Research, 32(2):441-447.

Loomis, John. 1996b. How Large is the Extent of the Market for Public Goods: Evidence from a Nationwide Contingent Valuation Survey. Applied Economics 28:779-782.

Loomis, John. 1999. Recreation and Passive Use Values from Removing Dams on the Lower Snake River to Increase Salmon. http://www.nww.usace.army.mil/lsr/REPORTS/misc_reports/passive.pdf

Loomis, J., T. Brown, B. Lucero and G. Peterson. 1997. Evaluating the Validity of the Dichotomous Choice Question Format. Environmental and Resource Economics 10:109-123.

Loomis, John and Pierre duVair. 1993. Evaluating the Effect of Alternative Risk Communication Devices on Willingness to Pay: Results from a Dichotomous Choice Contingent Valuation Experiment. Land Economics 69(3):287-298.

Loomis, John, Brian Roach, Frank Ward and Richard Ready. 1995. Testing Transferability of Recreation Demand Models Across Regions: A Study of Corps of Engineers Reservoirs. Water Resources Research 31(3):721-730.

Loomis, John and Richard Walsh. 1997. Recreation Economic Decisions, 2nd Edition. Venture Publishing, State College, PA.

Louviere, Jordan J. 1988. Analyzing decision making: metric conjoint analysis. Sage University Paper series on Quantitative Applications in the Social Sciences, series no. 07-067. Sage Publications, Beverly Hills, CA.

Machina, Mark. 1987. Choice Under Uncertainty: Problems Solved and Unsolved. Journal of Economic Perspectives 1:121-154.

Mackenzie, J. 1992. Evaluating recreation trip attributes and travel time via conjoint analysis. Journal of Leisure Research 24(2):171-184.

Mackenzie, J. 1993. A comparison of contingent preference models. American Journal of Agricultural Economics 75:593-603.

Magat, Wesley, W.Kip Viscusi and Joel Huber. 1988. Paired Comparison and Contingent Valuation Approaches to Morbidity Risk Valuation. Journal of Environmental Economics and Management 15(4):395-411.

Maler, K.G. Environmental Economics: A Theoretical Inquiry. Johns HopkinsUniversity Press, Baltimore, MD. 1974

McKean, John. 1981. An Input-Output Analysis of Sportsman Expenditures in Colorado. Colorado Water Resources Research Institute Technical Report #26. Colorado State University, Fort Collins, CO, January.

McKean, John. 1983. The Economic Base Model. Unpublished mimeo. Department of Economics, Colorado State University. Spring

Mendelsohn, R., J. Hof, G. Peterson and R. Johnson. 1992. Measuring Recreation Values with Multiple Destination Trips. American Journal of Agricultural Economics 74:926-933.

MIG, Inc. 1997. IMPLAN Professional. Minnesota IMPLAN Group. Stillwater, MN.

Mishan, E.J.,1981. Introduction to Normative Economics. Oxford University Press, Oxford, England.

Mitchell, R. and R. Carson. 1989. Using Surveys to Value Public Goods: The Contingent Valuation Method. Resources for the Future, Washington DC.

Mitchell, Robert and Richard Carson. 1985. Valuing Drinking Water Risk Reductions Using the Contingent Valuation Method. Draft Report to U.S. Environmental Protection Agency. Resources for the Future, Washington DC.

Natural Resources Defense Council. 2000. Arctic National Wildlife Refuge. http://www.nrdc.org/land/wilderness/arctic.asp.

The North American Commission for Environmental Cooperation. 2000. Taking Stock: North American Pollutant Releases and Transfers, 1997. May. http://www.cec.org/pubs_info_resources/publications/protect_human_enviro/tak97.cfm?varlan=english

Office of Management and Budget. 1992. Guidelines and Discount Rates for Benefit-Cost Analysis of Federal Programs. Circular A-94, Revised Transmittal Memo #64. Executive Office of the President, White House, Washington, D.C.

Palmer, Charles. 1983. I-O Study Team Contributions (originally published by Natural Resource Economics Division, USDA Economic Research Service, dated January 1976) reprinted in IMPLAN Training Notebook, Land Management Planning, Rocky Mountain Forest and Range Experiment, U.S. Forest Service, Fort Collins, CO.

Palmquist, Raymond. 1991. Hedonic Methods. In J. Braden and C. Kolstad, eds. Measuring the Demand for Environmental Quality. North Holland, New York.

Parsons, George R. and Aaron J. Wilson. 1997. Incidental and Joint Consumption in Recreation Demand. Agricultural and Resource Economics Review 26(1) (April):1-6.

Peterson, R.D. 1982. An Introduction to Input-Output Economics. Colorado Economic Issues Volume 6 (5). Cooperative Extension Service, Colorado State University, Fort Collins, CO. January.

Pope, C.A. 1989. Respiratory Disease Associated with Community Air Pollution and a Steel Mill, Utah Valley. American Journal of Public Health 79(5):623-628.

Powell, M. and J. Wilson. 1997. Risk Assessment for National Natural Resource Conservation Programs. Discussion Paper 97-49. Resources for the Future, Washington, D.C.

Power, Thomas. 1996. Lost Landscapes and Failed Economies. The Search for a Value of Place. Island Press, Covelo, CA.

Pyatt, Graham, and Jeffrey Round. 1985. Social Accounting Matrices: A Basis for Planning. The World Bank, Washington, D.C.

Quincy Library Group. 1993. 1993 Community Stability Proposal. August. http://www.qlg.org/pub/agree/comstab.htm

Quincy Library Group. 1998. Quincy Library Group Case Study. October. http://www.qlg.org/pub/miscdoc/overview.html

Radtke, Hans, Stan Detering, Ray Brokken. 1985. A Comparison of Economic Impact Estimates for Changes in the Federal Grazing Fee: Secondary vs Primary Data I/O Models. Western Journal of Agricultural Economics 10(2):382-390.

Reason Foundation. 1997. Rethinking EPA's Proposed Ozone and Particulate Standards. Policy Study #224. Executive Summary at http://reason.org/es224.htm.

Rice, D., T. Hodgson and A. Kopstein. 1985. The Economic Cost of Illness: A Replication and Update. Health Care Financing Reform 7(1):61-80.

Richardson, Harry. Input-Output and Regional Economics. John Wiley and Sons. New York. 1972.

Roe, B.; K.J. Boyle and M.F. Teisl. 1996. Using conjoint analysis to derive estimates of compensating variation. Journal of Environmental Economics and Management 31:145-159.

Rosenberger, R. and J. Loomis. 2000. Using Meta-analysis for Benefit-Transfer: In-sample Convergent Validity Tests of an Outdoor Recreation Database. Water Resources Research 36(4): 1097-1107.

Sassone, P. and W. Schaffer. 1978. Cost Benefit Analysis: A Handbook. Academic Press, New York, NY.

Sierra Club. 1996. The Morning NAFTA. The Planet 2(2) March. http://www.sierraclub.org/planet/199603/nafta.html

Smith, V. Kerry and William Desvousges. 1987. An Empirical Analysis of the Economic Value of Risk Changes. Journal of Political Economy, 95(February):89-114.

Smith, V. Kerry and Yoshiaki Karou. 1990. Signals or Noise? Explaining the Variation in Recreation Benefit Estimates. American Journal of Agricultural Economics 72:419-433.

Stevens, T. H., C. Barrett and C. E. Willis. 1997. Conjoint analysis of groundwater protection programs. Agricultural and Resource Economics Review 26(3):229-236.

Stokey, Edith and Richard Zeckhauser. 1978. A Primer for Policy Analysis. W.W. Norton Company, New York, NY.

Sullivan, Jay. 1983. Determining Changes in Final Demand for IMPLAN Economic Impact Analysis, in IMPLAN Training Notebook, Land Management Planning, Rocky Mountain Forest and Range Experiment, U.S. Forest Service, Fort Collins, CO.

Taylor, Michael L., Richard M. Adams, and Stanley F. Miller. 1992. Farm-Level Response to Agricultural Effluent Control Strategies: The Case of the Willamette Valley. Journal of Agricultural and Resource Economics 17:173-185.

The Economist. 1995. First, Ask the Locals. 337, November 4:28.

Tolley, G., D. Kenkel, and R. Fabian, eds. 1994. Valuing Health for Policy: An Economic Approach. University of Chicago Press.

U.S. Department of Agriculture. 1997a. Conservation Reserve Program Environmental Risk Assessment. Prepared by Farm Services Agency. Washington, D.C.

U.S. Department of Agriculture. 1997b. Environmental Quality Incentive Program Environmental Risk Assessment, Final. Prepared by Natural Resources Conservation Service. Washington, D.C.

U.S. Department of Commerce. 1986. Regional Multipliers: A User Handbook for the Regional Input-Output Modelling System (RIMS II). Bureau of Economic Analysis. U.S. Department of Commerce, Washington DC.

U.S. Department of Commerce. 1997. Regional Multipliers: A User Handbook for the Regional Input-Output Modelling System (RIMS II). Third Edition. Bureau of Economic Analysis. U.S. Department of Commerce, Washington DC. http://www.bea.doc.gov/bea/ARTICLES/REGIONAL/PERSINC/Meth/rims2.pdf

U.S. Department of Interior. 1994. Natural Resource Damage Assessments; Final Rule. Federal Register 59(58) Pages?. Washington DC.

U.S. District Court of Appeals. 1989. State of Ohio v. U.S. Department of Interior. Case No. 86-1575. District of Columbia. July 14.

U.S. Environmental Protection Agency. 1992. Framework for Ecological Risk Assessment. EAP/630-r-92-0001. Washington, D.C. Februrary.

U.S. Environmental Protection Agency. 1996. Proposed Guidelines for Ecological Risk Assessment. EAP/630-R-95/002B. Washington, D.C. August.

U.S. Environmental Protection Agency. 1998a. Draft Inventory of US Greenhouse Gas Emissions and Sinks (1990-1996). EPA236-R-98-006, March. http://www.epa.gov/globalwarming/publications/emissions/us1998/index.html.

U.S. Environmental Protection Agency. 1998b. Economic Analysis of Toxic Substances Control Act Section 403: Hazard Standards. Economic and Policy Analysis Branch, Office of Pollution Prevention and Toxics, Washington, D.C.

U.S. Environmental Protection Agency. 1998c. Emissions of Nitrous Oxide from Highway Mobile Sources: Comments on the Draft Inventory of U.S. Greenhouse Gas Emissions and Sinks, 1990-1996, EPA420-R-98-009, August, http://www.epa.gov/otaq/climate.htm, p. 4.

U.S. Geological Survey. 1998. "New Estimates of Petroleum Resources of the Arctic National Wildlife Refuge 1002 Area," May 17. Available at http://www.usgs.gov/public/press/public_affairs/press_releases/pr517.htm

U.S. Global Change Research Information Office. 2001. "Are Human Activities Contributing to Climate Change?" Available at http://www.gcrio.org/ipcc/qa/03.html.

U.S. Water Resources. 1979. Procedures for Evaluation of National Economic Development (NED): Benefits and Costs of Water Resources Planning (Level C) Final Rule. Federal Register 44(242): 72892-977.

U.S. Water Resources Council. 1983. Economic and Environmental Principles and Guidelines for Water and Related Land Resources Implementation Studies. U.S. Government Printing Office, Washington DC. March 10

Vartia, Yrji. 1983. Efficient Methods of Measure Welfare Change and Compensated Income in Terms of Ordinary Demand Functions. Econometrica 51(1):79-98.

Viscusi, Kip. 1993.The Value of Risks to Life and Health. Journal of Economic Literature 31(4): 1912-1946.

Wald, Matthew. 1998. Autos' Converters Cut Smog But Add to Global Warming. The New York Times May 29, page A1, Column 4.

Walsh, R., J. Loomis and R. Gillman. 1984. Valuing Option, Existence and Bequest Demands for Wilderness. Land Economics 60(1): 14-29.

Walsh, R. G., D. M. Johnson and J. R. McKean. 1992. Benefit Transfer of Outdoor Recreation Demand Studies: 1968-1988. Water Resources Research 28(3): 707-713.

Ward, Frank and John Loomis. 1986. The Travel Cost Demand Model as an Environmental Policy Assessment Tool: A Review of the Literature. Western Journal of Agricultural Economics 11(2): 164-178.

Waters, Edward, Bruce Weber and David Holland. 1999. The Role of Agriculture in Oregon's Economic Base: Findings from a Social Accounting Matrix. Journal of Agricultural and Resource Economics 24(1): 266-280.

Weitzman, Martin. 1998. Why the Far-Distant Future Should be Discounted at Its Lowest Possible Rate. Journal of Environmental Economics and Management 36: 201-208.

Weitzman, Martin. 2001. Gamma Discounting. American Economic Review 91(1): 260-271.

Welsh, Michael. 1986. Exploring the Accuracy of the Contingent Valuation Method: Comparisons with Simulated Markets. Ph.D. dissertation. Department of Agricultural Economics, University of Wisconsin, Madison, WI.

Young, Robert. 1973. The Price Elasticity of Demand For Water: A Case Study of Tucson, Arizona. Water Resources Research 9(6): 1068-1072.

Zeckhauser, Richard and W. Kip. Viscusi. 1990. Risk within Reason. Science 248(May) :559-564.

Zinkhan. C. F., T. P. Holmes and D. E. Mercer. 1994. Conjoint analysis: a pragmatic approach for the accounting of multiple benefits in southern forest management. Southeastern Center for Forest Economics Research, Research Triangle Park, NC. SCFER Working Paper No. 76, 16 pp.

Index

MIX
Papier aus verantwortungsvollen Quellen
Paper from responsible sources
FSC® C105338

If you have any concerns about our products,
you can contact us on
ProductSafety@springernature.com

In case Publisher is established outside the EU,
the EU authorized representative is:
Springer Nature Customer Service Center GmbH
Europaplatz 3, 69115 Heidelberg, Germany

Printed by Libri Plureos GmbH
in Hamburg, Germany